An Introductory Course in Practical Organic Chemistry

An Introductory Course in Practical Organic Chemistry

F. D. GUNSTONE, D. M. SMITH
University of St. Andrews
J. T. SHARP
University of Edinburgh

With a foreword by
J. I. G. CADOGAN

METHUEN & CO LTD
11 NEW FETTER LANE, LONDON EC4

First published 1970

Printed in Great Britain by
William Clowes & Co Ltd
London & Beccles

SBN 416 13940 X

Distributed in the U.S.A. by
Barnes & Noble, Inc.

Foreword

by J. I. G. Cadogan

Forbes Professor of Chemistry in the University of Edinburgh, formerly Purdie Professor of Chemistry in the University of St. Andrews.

The arrival of new techniques and new instrumentation which has transformed organic chemistry during the last decade has put new demands on the teaching of the experimental branch of the subject. It is no longer sufficient to teach our students how to crystallize, distil and separate materials by classical methods alone, in view of the powerful chromatographic and spectroscopic aids now in every day use in chemistry. We must now provide instruction in all these methods and there is a strong case to be made for introducing students to the more sophisticated tools as soon as possible in their University careers. This was the philosophy behind the practical instruction introduced in St. Andrews when the authors were all colleagues there.

This compact book sets out, very clearly, the details of the resulting two-year introductory course, which contained a sufficient variety of experiments to cope with the fairly wide range of practical attainment (ranging from almost nothing to Scholarship level) already achieved by the various students entering the first year in a Scottish University. The rationale of the course has already been touched on; this is further elaborated by the authors in their preface and in the introduction to the 'Identification' chapter on page 143. Features of particular value or novelty include the chapters on spectroscopy (5 and 6), Experiments 11 (p. 69) and 22 (p. 89), which rather nicely bring out mechanistic points, while providing experience in gas-liquid chromatography, Experiment 48 (p. 122), which illustrates substituent effects in a kinetic study of the hydrolysis of aroyl chlorides, and Experiments 49 (p. 124) and 50 (p. 130), which provide a simple and straightforward introduction to the application of n.m.r. spectroscopy in organic chemistry. In addition the classical techniques which are still, and will continue to be, the bedrock of organic chemistry are well represented.

The course on which this book is based is well tried, has been well received in practice, and provides a good foundation on which to graft more advanced experiments for the senior student. As such it will be very valuable to teachers and to the taught.

J. I. G. Cadogan
March 1970

Authors' preface

This book is based on the courses in practical organic chemistry given in the University of St. Andrews to our first and second year classes. All science students taking chemistry, and all medical students take the first year course, which consists of nine classes each of four hours. About 100 students attend each laboratory class, and during the course they undertake about half of the experiments described here. The second year class, which includes Ordinary Degree students as well as potential honours students in Chemistry, Biochemistry, and Geology, is taught in groups of about 30. During their course of 16 meetings of three hours, the remaining experiments are carried out and the identification of unknown compounds is undertaken.

We believe that this course has certain novel features which warrant its being offered to a wider public, and we draw special attention to four of these.

1. In recent years great changes have occurred in the way in which chemistry is presented in the lecture theatre, and a mechanistic approach of some sort is now followed almost universally. These changes have not always been reflected in the associated laboratory classes (as students are quick to point out), and we have had this in mind in devising this course. Our experiments are arranged and designed to emphasize points of mechanism as well as to illustrate and teach techniques. Whilst not neglecting older classical procedures, we have purposely introduced techniques of chromatography and spectroscopy from the beginning of our introductory course. They are an integral and important part of present-day chemistry, and though developed later in time they are not more difficult to use. (Crystallization is one of the oldest techniques in chemistry, yet one of the most difficult to do properly.) We do not subscribe to the view that the student should be trained on a shortened historical scale with classical methods for first year students and modern methods reserved for final year students. Sophisticated instruments are now an important feature of every science department and we use these in our elementary courses as much as we can. We

have not given details about their use since these will depend on the model employed. In general, our undergraduates use the same type of instrument as is used in our research laboratories and not a cheaper, simpler, or inferior model. Although a little more expensive initially, this permits more interchangeability and a more economic use of instruments.

2. We are aware of the changes occurring in methods of training at school, and believe that the open-ended nature of many of our experiments will be in line with the deductive approach which is familiar to many students when they come to university. Many of our experiments fall into related groups and it is expected that students will carry out one such experiment and then compare and discuss their results with other students. This allows students to share one another's practical experience, so making more profitable use of laboratory time.

3. From the beginning of the course we make extensive use of column chromatography, thin-layer chromatography, and gas liquid chromatography. This last chromatographic technique in particular, with its quantitative possibilities, introduces a new dimension into the teaching of practical organic chemistry. Students become much more critical about the purity of their products and, more important, it is possible to carry out new types of experiments which produce mixtures of compounds. The results of such experiments lead to many interesting discussions on theoretical points.

4. Spectroscopic techniques have greatly changed the methods by which chemists identify unknown compounds, but these developments are not always reflected in the procedures which elementary students are required to follow. We use infrared spectroscopy from the very beginning of our practical course, and several of the early experiments are designed to provide practice in the interpretation of infrared spectra. The consequences of this are seen in our procedures for qualitative analysis, and our philosophy about this is fully described on p. 143. We have also included some experiments on the interpretation of simple n.m.r. spectra.

We hope this course will be a valuable introduction to practical organic chemistry for students in universities and technical colleges, and that it may find some use in schools. It is not essential to follow the order we have set out here, but early experiments have been described in much greater detail than those that follow. Some techniques in practical chemistry are easy to demonstrate but more difficult to describe, and since we do not expect students to undertake such a course without a teacher, we have made frequent references to the need to consult an instructor.

We acknowledge the help received from many of our colleagues in several ways. Dr. A. R. Butler provided the kinetic experiment (number 48). Dr. R. K. Mackie and Mr. A. Watson helped us with the section on n.m.r. spectroscopy. Mrs. P. A. Sugden and Miss H. Clacher, with their considerable experience of first year students, helped us to match our ideas with the ability (and limitations) of such students with wide-ranging interests. We thank Professor J. I. G. Cadogan for asking us to arrange this course and leaving us to get on with it!

St. Andrews,
July 1969

F.D.G., J.T.S., D.M.S.

Contents

List of experiments and techniques used

The following techniques are described and used in the experiments indicated below. An asterisk indicates that not all references to the technique are included but only the major ones where more detailed instructions are given. Attention is drawn to the explanation and general instructions given in Part One.

LABORATORY ACCIDENTS

Safety spectacles should be worn at all times in a chemical laboratory.

Familiarity with practical chemistry should not lead to a contempt for its hazards. Many organic chemicals, particularly solvents, are both volatile and inflammable. Since most chemicals are toxic to some degree they should be kept off the skin.

FIRE Know the location of your nearest extinguisher and fire-blanket.

Attempt to contain the fire by using a carbon dioxide extinguisher or sand (NOT WATER). Do not take personal risks; evacuate the area if necessary.

Burning clothes. Smother the flames with a fire-blanket (or lab. coat) or in extreme cases use a carbon dioxide extinguisher.

Burns. Drench with cold water and obtain medical attention.

CHEMICAL HAZARDS

Immediate first-aid measures are outlined below. In all but trivial accidents skilled First Aid or medical attention should then be sought.

Eyes. Irrigate thoroughly with water using an eye wash bottle if available.

Mouth. Wash out thoroughly with water.

Ingestion. Drink plenty of water as a diluent.

Skin. Drench with cold water, then wash with soap and cold water for water-insoluble organic compounds.

Lungs. The affected person should be removed from the contaminated area, kept warm, and made to rest. If breathing has stopped, use artificial respiration.

PART ONE

1. Crystallization

Crystallization is an important procedure used to purify solids by removing impurities of differing solubility. Briefly the method consists of (1) dissolving the solid in hot solvent, (2) filtering the hot solution to remove insoluble impurities, if these are present, (3) cooling the filtrate until crystallization is complete, (4) separating crystals from mother liquor by filtration, and (5) drying the crystals. Dissolving a solid in a solvent and then evaporating all the solvent may produce crystals but it will not purify the compound since no impurities have been removed.

Purification may not be complete after one crystallization *and the process may have to be repeated*, sometimes from a different solvent. Purification should be combined with maximum recovery of product.

1. Equipment. The apparatus used must be commensurate in size with the amount of material to be handled. Crystallization should be carried out in conical flasks or tubes which can be closed. Beakers are not suitable vessels for use with organic solvents which are frequently volatile and inflammable. Filtration is normally effected under reduced pressure using a Büchner or Hirsch funnel (Figure 1), fitted with a filter paper or having a sintered plate, along with a suitable filter tube or flask.

2. Procedure. Perhaps the most difficult part of crystallization is the selection of the solvent. Beginners will almost certainly be told what solvent to use and instructions on selecting a solvent are not included at this stage but are described on p. 5.

The solid (previously crushed if the crystals are very hard or very large) is placed in a conical flask along with a few anti-bumping granules and what is judged to be an insufficient quantity of solvent for complete solution. After fitting a reflux condenser, the mixture is heated to boiling, with occasional swirling, on a water bath or steam bath if possible. More solvent is added in small portions if required and refluxing continued until solution seems to be complete*.

The hot solution need be filtered *only if insoluble impurities are present.* It is desirable to use a warm funnel to prevent the solution cooling so much that crystals are lost on or in the funnel, but if the funnel is too hot there may be excessive loss of solvent under the reduced pressure, possibly with vigorous boiling, and deposition of crystals on the inside of the porcelain funnel. It is often convenient to collect the clear warm filtrate in a test tube contained in the filter tube (Figure 1). After filtering the solution, the funnel should be rinsed with a little warm solvent. The tube containing

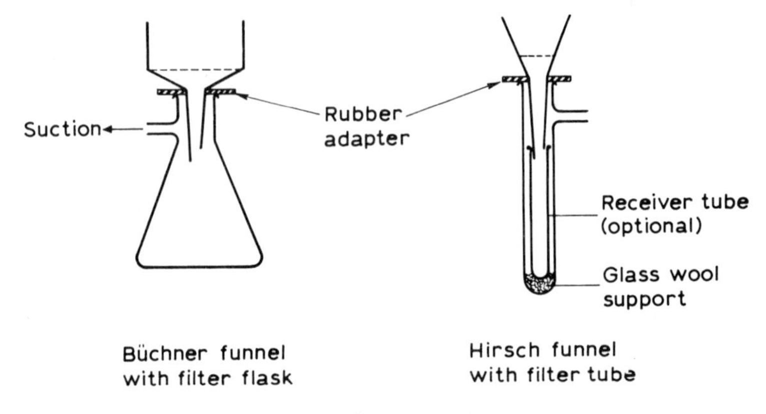

Figure 1. *Filtration equipment*

the filtrate is then removed and warmed to redissolve any crystals which may have appeared.

When the solution cools it should deposit crystals of the solute being purified. If this separates as a liquid which subsequently solidifies, either more solvent must be used so that separation does not occur until a lower temperature, or the solution must be seeded with some of the crystalline material at the first sign of oil-formation, or the solution must be allowed to cool very slowly. If crystals do not appear on cooling, crystallization may be induced by scratching the inner walls of the vessel with a glass

* If highly coloured or oily impurities are present the addition of a little dried finely-powdered charcoal will assist their removal. The solution must be cooled before adding the charcoal and then re-heated. The hot suspension is filtered through a filter paper covered with a thin layer (3-5 mm) of celite which has been moistened with warm crystallization solvent.

rod or by addition of a crystal of the substance being purified. If both these steps fail, some of the solvent may have to be distilled off. Complete crystallization may take several hours and the crystallizing solution should be left as long as convenient before filtering.

The crystals are collected by suction filtration using a Büchner or Hirsch funnel of appropriate size fitted to a dry filter tube or flask (Figure 1) and the crystals are pressed down with a spatula. They must be washed by releasing the suction, adding just enough solvent to wash the crystals thoroughly, and by sucking them dry as quickly as possible. A second crop of crystals can sometimes be obtained by concentrating the mother liquor but their purity must be carefully checked. *The filtrate must never be discarded until all the required material has been recovered.*

Crystals are freed from adhering solvent by spreading them on a watchglass. Volatile solvents quickly evaporate. A vacuum desiccator, or an oven kept well below the melting or sublimation temperature of the crystalline product, may be required with non-volatile solvents. Finally, the melting point of the crystals is determined.

A special procedure with small amounts of material is described in section 4.

3. Choice of solvent. A little of the solid (*ca.* 50 mg) is placed in a clean test tube (75 x 10 mm) along with three or four drops of solvent, and the mixture shaken or stirred with a glass rod. The solvent is unsuitable if the solid dissolves at room temperature. Otherwise, the mixture is heated until all or most of the solid has dissolved, freed from any insoluble material by decantation, and cooled to see if crystallization occurs.

This process is repeated with several solvents until the most satisfactory one has been found. It is usual to try water (seldom successful), methanol, ethanol, acetic acid, benzene, petroleum (available in several boiling ranges) and, less commonly, ether, chloroform, ethyl acetate, methyl cyanide, and dimethylformamide.*

If no single solvent is appropriate, mixed solvents may be necessary. These include aqueous alcohol, aqueous acetic acid, or benzene-petroleum mixtures. For example, a substance which is very soluble in alcohol and almost insoluble in water may crystallize well from a mixture. The correct procedure is to dissolve the solid in a little

* All these organic solvents have their individual hazards and contact with both liquid and vapour should be avoided. Special attention is drawn to benzene which is now recognized as a cumulative poison affecting the blood. Even low concentrations of benzene vapour should therefore be avoided.

boiling alcohol and add warm water dropwise. Each drop will produce a cloudiness which at first clears on mixing. When the warmed solution just fails to clear on shaking, a few drops of alcohol are added, the mixture is re-heated, and set aside for crystallization. Any insoluble impurities are best removed by filtering the hot alcohol solution before adding water.

*4. Crystallization of small amounts of material in Craig tubes.** Craig tubes (**A**, Figure 2) are particularly useful when the amount of material to be crystallized is very small. Their use reduces the number of items used and the amount of handling required. Used in conjunction with the filtration device (**B**) they are particularly useful for preparing pure samples for quantitative analysis, but the Craig tube can be used without this additional device.

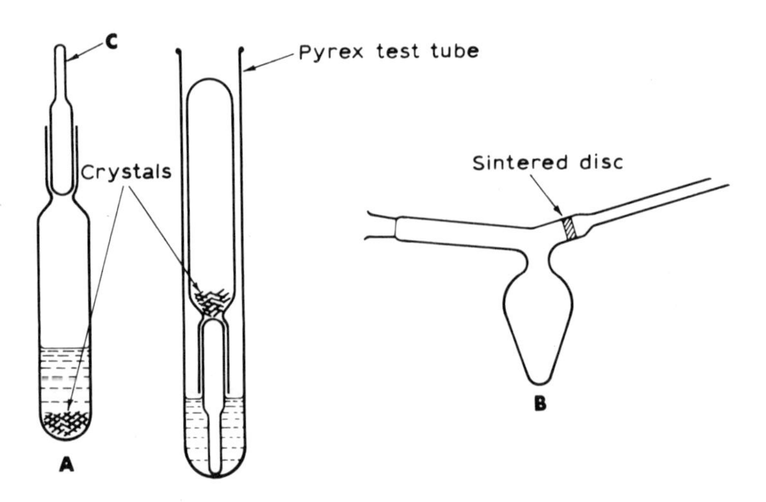

Figure 2. *Filtration in a Craig tube*

The sample is introduced into **B** through the side-arm followed by the crystallizing solvent. The bulb is then heated to dissolve the solute and to warm the sintered disc and spout of **B** by means of hot vapour. With a rubber bulb attached to the side-arm the hot solution is forced through the sintered disc into a Craig tube which is then

* M. Martin-Smith, *Laboratory Practice* (1958), 572.

closed with a loosely fitting stopper **(C)**. After the sample has crystallized, the Craig tube and stopper are inverted in a Pyrex test tube and the whole unit centrifuged for a few minutes. This removes the mother liquor into the test tube and leaves the crystals in the Craig tube where they can be vacuum dried.

2. Melting point

The melting point (m.p.) is an important property of an organic solid. It is a means of identification and a measure of purity, and *the preparation or purification of a solid is not complete until this value has been determined and recorded.*

Although the term melting point is in common use, very few substances melt instantaneously and the melting range (e.g. 120-122°) should be quoted. This indicates that melting was first observed at 120° and was complete at 122°. Pure compounds have a very short melting range; impure compounds have depressed melting points and bigger melting ranges.

Since melting points are normally depressed by impurities, it is possible to prove or disprove the identity of two solids of similar melting point by a mixed melting point determination. From the figures given below, A and B are the same compound but C and D are different. The mixture of C and D has a lower melting point and wider melting range than either of the single substances.

	m.p.	mixed m.p.
A	120-121°	119-121°
B	119-121°	
C	130-132°	123-128°
D	131-133°	

When carrying out a mixed melting point determination the melting point of each substance and of an approximately 1:1 mixture should be measured. The two substances must be thoroughly ground together for this purpose.

Your instructor will show you how to determine a melting point using the apparatus available in your laboratory.

3. Distillation

Distillation is an important method of purifying liquids based on differences in boiling point.

Simple distillation

1. Equipment. The apparatus required for distillation includes a distillation flask, a distillation head with thermometer, a condenser, a receiver-adapter, a receiver, and an appropriate source of heat (Figure 3). This simple assembly is not very efficient and will only separate liquids of very different boiling points. It could be used, for

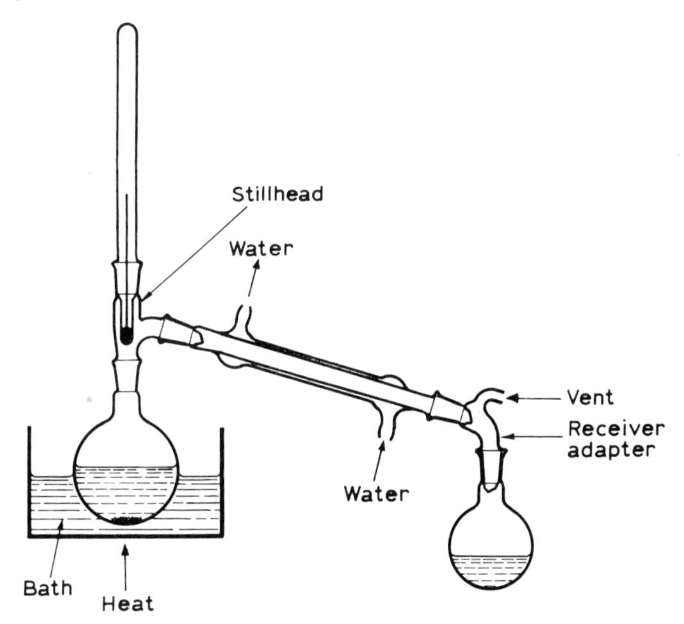

Figure 3. *Distillation at ordinary pressure*

example, to separate ether or petroleum (b.p. 40-60°) from a reaction product distilling above 100°.

The distillation flask should be round-bottomed or pear-shaped and never conical or flat-bottomed. It should be of such a size that it is one-third to two-thirds full at the beginning of the distillation. The thermometer is fitted into the stillhead so that its bulb is just below the level of the side-arm (Figure 3).

A water condenser is used for liquids distilling below about 150°. An air condenser will suffice above this temperature, or the water can be run out of the water condenser. Special care must be taken to see that the condenser is working efficiently when distilling volatile and inflammable liquids such as ether, petroleum, or benzene.

The receiver-adapter must have a vent open to the atmosphere. *Distillation must never be conducted in a closed system.* When it is essential to exclude moisture a loosely packed drying tube can be attached to the vent. With highly inflammable liquids it is desirable to convey uncondensed vapours away from any flame by attaching a length of rubber tubing to the vent and allowing the end to hang well below bench level.

A flask of appropriate size is used as receiver. This should be preweighed when used to collect a reaction product.

The distillation flask may be heated directly with a Bunsen flame only when aqueous solutions are being distilled or in exceptional circumstances. More generally, liquids boiling below about 80° are heated on a water bath or steam bath, and an oil bath is used for liquids boiling between 80 and 180°. A free flame or metal bath is sometimes used above this temperature but it is preferable to distil high-boiling liquids at reduced pressure. A thermometer should always be placed in an oil bath to allow careful temperature control. The temperature in the bath is usually 10-30° higher than that of the liquid in the flask and 30-50° higher than the distillation temperature. An oil bath should not be used above 220° and even below this temperature the distillation should be carried out in a fume cupboard if possible. With high-boiling liquids the stillhead may have to be insulated with asbestos string.

When distillation is undertaken for the first time the apparatus should be checked by the instructor before use.

2. *Distillation procedure.* A few anti-bumping granules or porous chips are added to the distillation flask before heating. If distillation has to be interrupted, more

anti-bumping granules should be added before resuming. *Solid material should never be added to a hot liquid which may be near to its boiling point.*

The flask is heated gently and, when distillation starts, the rate of heating is adjusted so that distillation proceeds slowly but steadily. The boiling range of a product should generally not exceed 5° and should always be recorded.

Steam distillation

It is often convenient to separate organic compounds from non-volatile organic or inorganic contaminants by co-distillation with steam. This can be done by passing

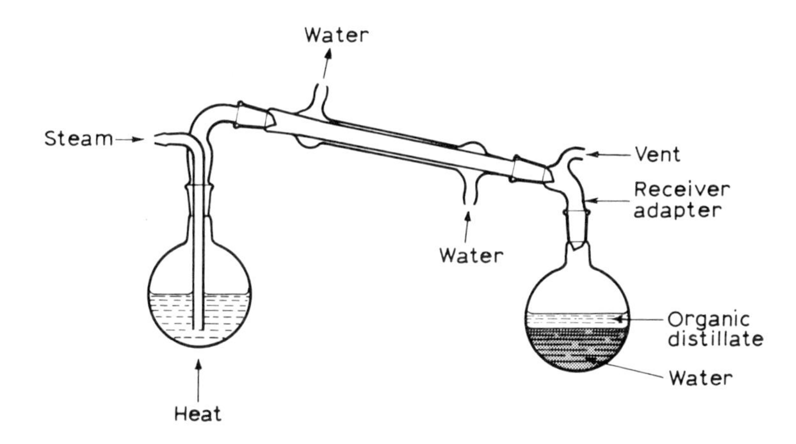

Figure 4. *Steam distillation*

steam into a heated flask containing the liquid to be distilled (Figure 4). The mixture of steam and organic compound that distils out is condensed and collected in the usual way and the components subsequently separated from each other.

On a small scale it is often convenient to generate the steam *in situ* by boiling an aqueous solution or suspension (see for example, Experiments 16, 40, 41, 43 and 44). Additional water may have to be added during the distillation to prevent the solution becoming too concentrated.

Distillation under reduced pressure (vacuum distillation)

1. Equipment. High-boiling liquids or those which decompose at or below their normal boiling point are generally distilled at lower temperatures under reduced pressure. With a water pump a pressure of 10-20 mm can be obtained and under these conditions boiling points are reduced by about 100°. Lower pressures (*ca.* 0.1 mm) may be obtained with oil pumps.

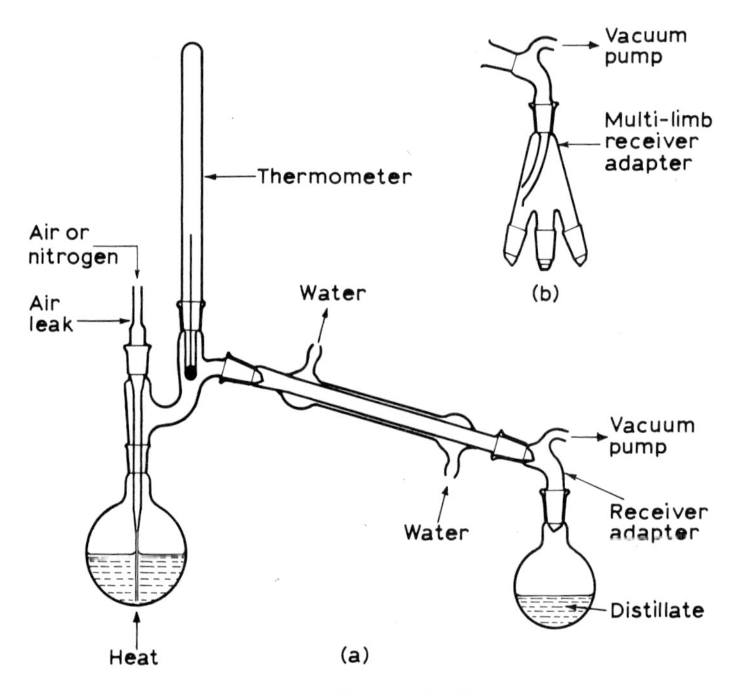

Figure 5. *Vacuum distillation*

In vacuum distillation (Figure 5*a*) the distillation flask should only be one quarter to one half full and the simple stillhead is normally replaced by a Claisen stillhead. This has two necks; one carries the thermometer and the other is usually fitted with a capillary leak. The receiver-adapter is connected to a manometer, trap, and pump. Since the pressure should be kept constant during distillation it is advisable to use equipment which allows more than one fraction to be collected without disturbing the vacuum (Figure 5*b*). A heating bath is essential for vacuum distillation: a Bunsen flame should never be used.

Anti-bumping granules will not prevent 'bumping' in a reduced pressure distillation and instead the flask should be packed with glass wool or, preferably, equipped with a capillary leak. This must be exceedingly fine and extend to within 2-3 mm of the

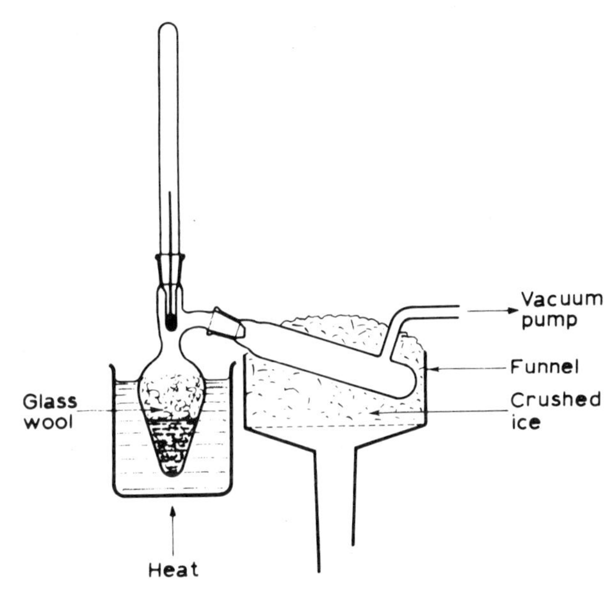

Figure 6. *Vacuum distillation of low-melting solids*

bottom of the flask. When the pressure is reduced in the distillation system, a stream of small bubbles will be sucked through this capillary. Dry nitrogen should be used instead of air whenever the compounds are oxidizable by atmospheric oxygen.

A simple apparatus suitable for the distillation of small amounts of low-melting solids is shown in Figure 6.

2. *Procedure.* Before distillation starts, the liquid must be at room temperature and be free of any volatile material such as ether, otherwise uncontrollable boiling will occur when the pressure is reduced. The water pump should be fully on throughout

the distillation, and the pump and distillation apparatus should be connected through a three-way tap. A satisfactory vacuum must be attained before heating the distillation flask. The glass joints may be greased sparingly at the top if this is necessary. When collecting the distillate, the distillation temperature and pressure must be recorded. The source of heat is removed and the apparatus cooled before releasing the vacuum. *This is not released by turning off the pump but by opening another tap in the system.*

4. Chromatography*

The separation of complex mixtures and the isolation of individual components is important in all areas of organic chemistry. The classical methods of crystallization, distillation, and the separation of acidic, basic, and neutral compounds are now complemented by several chromatographic techniques. These last are often capable of effecting otherwise difficult separations.

All chromatographic procedures involve the interaction of a *mobile* phase, which may be a gas or a liquid, and a *stationary* phase, which may be liquid or solid. When the stationary phase is solid, the basis of separation is adsorption, and when the stationary phase is liquid, separation is based on partition. Some of the common forms of chromatography are set out in the following table.

Type of chromatography	*Mobile and stationary phase*	*Principle of separation*	*Uses*
Column	Liquid/solid	Adsorption	Preparative scale separations
Thin layer (TLC)	Liquid/solid	Adsorption	Qualitative analysis and small-scale preparative separations
Gas liquid (GLC)	Gas/liquid	Partition	Qualitative and quantitative analysis; also preparative scale separations
Paper	Liquid/liquid	Partition	Qualitative and quantitative analysis of polar organic and inorganic compounds

Column chromatography

Column chromatography dates back to the late nineteenth century but was not used extensively until the 1930s. The name chromatography derives from its initial use to separate coloured compounds but its value is in no way restricted to such compounds.

* Useful supporting film: Chromatography (ICI Ltd.)

The stationary phase is a powdered adsorbent such as alumina (Al_2O_3) or silica gel (SiO_2). This is packed into a column (Figure 7*a*) together with an organic solvent such as petroleum or benzene which is to be used as the mobile phase.

The mixture to be separated is applied to the top of the column where it is adsorbed by the stationary phase, and the eluting solvent (mobile phase) is then passed continuously through the column (Figure 7*b*). Each component of a mixture is carried down the column by the mobile phase at a speed dependent on its affinity for the

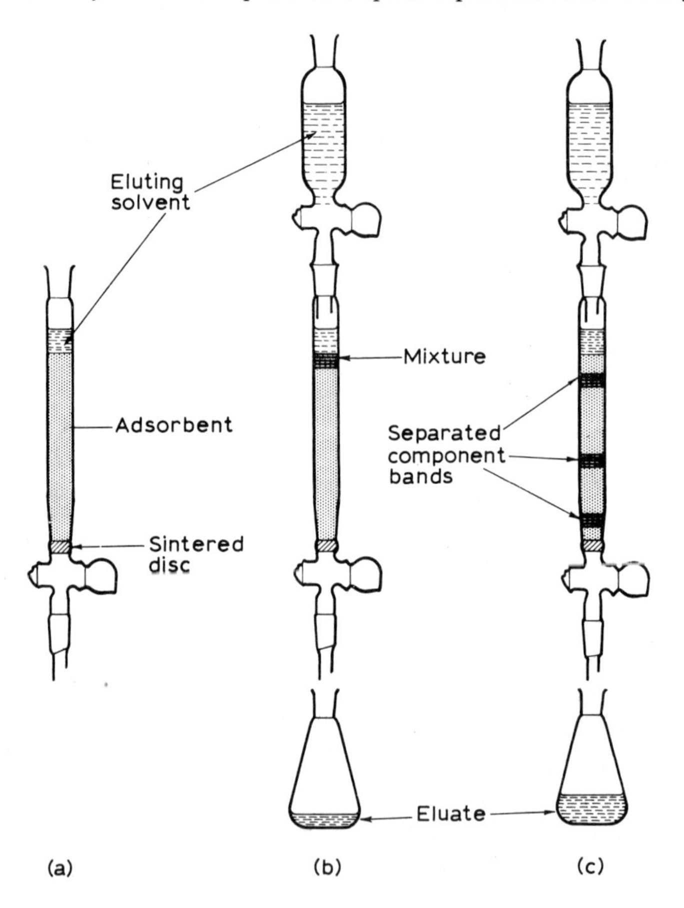

Figure 7. *Column chromatography*

adsorbent. This process is described as developing the column. Ideally the mixture will separate into a number of discrete bands (Figure 7*c*) and elution is continued until all components have been washed through the column. Polar compounds such as alcohols (ROH), amines (RNH_2), or carboxylic acids ($R.CO_2H$) are strongly adsorbed and move slowly or not at all; less polar compounds such as hydrocarbons (RH), ethers (R_2O), and alkyl halides (RX) are adsorbed less strongly and move more quickly.

The separation of coloured compounds is readily observed. With colourless compounds it is usual to collect the eluate in batches, distil off the eluting solvent, and examine the residue in some appropriate way. Thin-layer and gas liquid chromatography are particularly suitable for this purpose. Special methods for continuously monitoring eluates have also been described.

Compounds differing markedly in polarity, such as a hydrocarbon and an alcohol, are easily separated on a short column but compounds more alike in polarity may require longer columns and a careful selection of adsorbent and eluting solvent if they are to be cleanly separated.

1. Column packing. The amount of adsorbent required is usually 30-100 times the amount of material to be separated. Long thin columns give better separation than short wide columns.

The usual way to pack a column is to half-fill it with the eluting solvent and to pour slowly into this a mobile slurry of the adsorbent and eluting solvent. The column is tapped gently with a piece of rubber tubing and solvent is allowed to run slowly from the column as the adsorbent settles. Columns should be packed as evenly as possible: unevenly packed columns give poor separation. The column must not be allowed to run dry until the desired separation has been achieved.

2. Application of the sample. Liquid mixtures may be applied directly to the column but solids are applied as concentrated solutions in the developing solvent or, if this is not possible, in a solvent of as low an eluting power as possible (see below).

The solvent used to prepare the column is run off until there is only a small layer (*ca.* 3 mm) on top of the adsorbent. The sample is then added using a pipette and more solvent is drained away until the liquid level is again just above the adsorbent. This process is repeated with some of the eluting solvent (*ca.* 1 ml) and elution continued with larger volumes of solvent.

3. *Eluting solvents.* Chromatographic separations are usually effected with a range of solvents of increasing eluting power (see below). A non-polar solvent is used first to elute the less strongly adsorbed components of the mixture and this is replaced progressively by solvents of greater eluting power. The most commonly used solvents* are set out below:

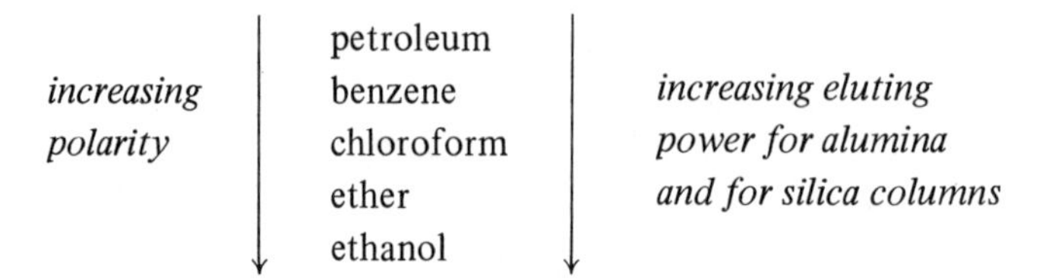

Thin-layer chromatography (TLC)

Thin-layer chromatography is a recent development used extensively for qualitative analysis and for preparative separation on a small scale. TLC usually gives better separation of the components of a mixture than does column chromatography and has the advantage of being a fairly quick process.

The adsorbent is most commonly a thin layer of alumina or silica gel containing a little calcium sulphate to increase the strength of the layer. Its thickness is generally about 0.25 mm for analytical purposes and up to 1 mm for preparative purposes. The layer is coated on to glass plates (usually 5 x 20 cm or 20 x 20 cm) but microscope slides (25 x 75 mm, Figure 8*a*) are also recommended. These small plates are easy to coat with adsorbent, quick to use, and useful for demonstrating TLC techniques. They are frequently used for monitoring reactions and column chromatograms.

The mixture to be separated, dissolved in an appropriate solvent, is spotted on to the plate (Figure 8*b*) and after the solvent has evaporated the plate is placed in a developing jar (Figure 8*c*) containing a little of the developing solvent. Solvent rises through the adsorbent layer by capillary attraction, and the various components of the mixture ascend at different rates depending on their differing affinity for the adsorbent. When the solvent front has almost reached the top of the adsorbent layer, the components should ideally be well separated. The distance travelled by the components can be increased by using a solvent of higher polarity and vice versa (see

* See footnote on p. 5

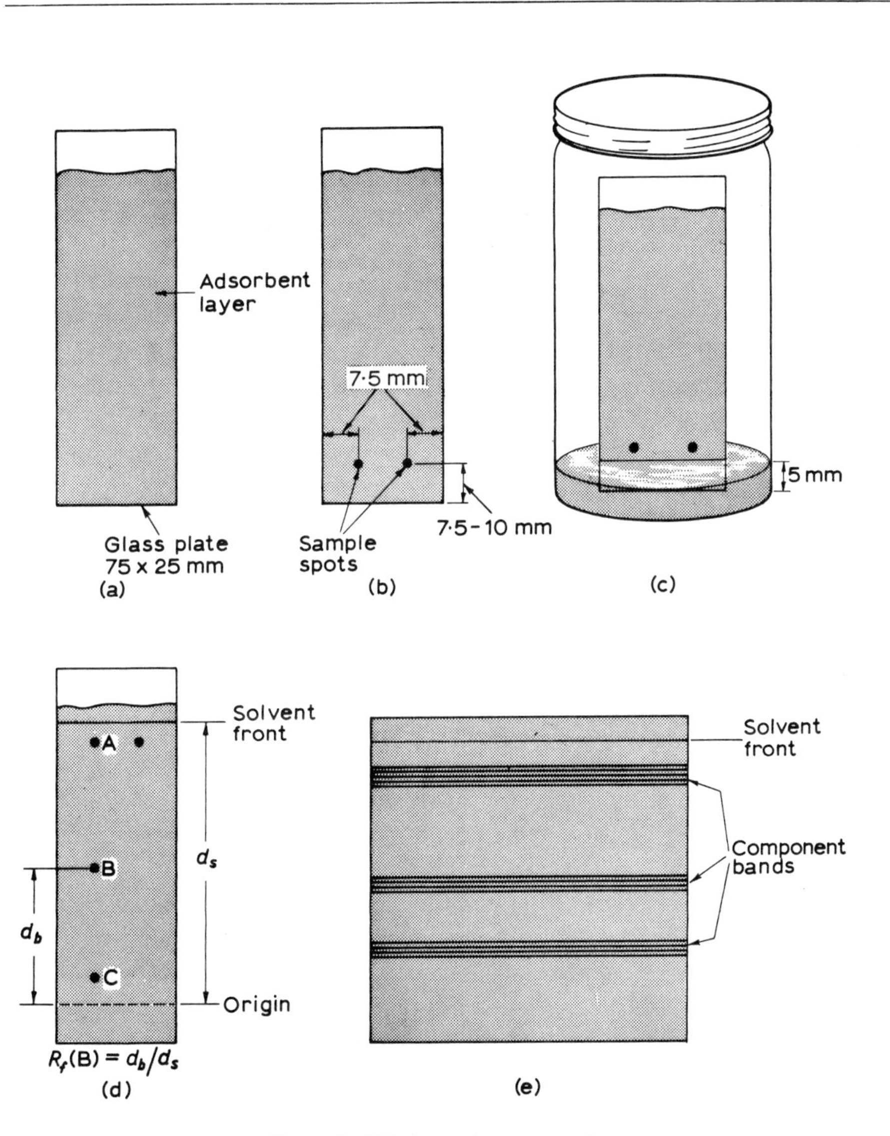

Figure 8. *Thin-layer chromatography*

the list of solvents given on p. 18). A developed plate should look something like Figure 8*d*.

The R_f value (see Figure 8*d*) provides a useful index for comparing two compounds but these values are not always very reproducible in TLC. When trying to prove the identity of two compounds by TLC the unknown and an authentic sample should be run side by side on the same plate. Identity of chromatographic behaviour is then indicative of identity of structure but does not furnish final proof of this.

In preparative TLC a larger amount of material (20-200 mg, depending on the ease of separation) is applied in a narrow band to a thicker layer (1 mm) of adsorbent on a square plate (20 x 20 cm). The developed chromatogram should look something like Figure 8*e*. The adsorbent is scraped from the plate in bands and the separated components are extracted from the adsorbent with a suitable solvent.

1. Coating the plates. The plates must be coated with a uniform layer of adsorbent and this is achieved for larger plates with a commercial spreading device.

Microscope slides can be coated by dipping them in pairs, held back to back, into a 1:1 slurry of adsorbent and methanol, quickly removing them, and wiping the long edges with the thumb and forefinger. The coated plates are separated, placed horizontally on a suitable rack, and activated by drying at 120° for 30 minutes. They can be stored in a desiccator or drying cabinet. The thickness of the layer made in this simple way depends on the composition of the slurry. A 1:1 mixture of adsorbent and solvent gives a reasonable layer on microscope slides.

2. Sample application. Samples are normally applied as 1-2% solutions from a capillary dropper. Overloading gives rise to unduly large spots or extensive streaking after development. Suitable dropping-tubes are easily made from a melting-point tube by drawing out the middle, using a microburner, and breaking the tube into two equal parts. The dropper is charged by capillary attraction and will be discharged when the liquid on the tip of the dropper touches the adsorbent surface. The spots should be as small as possible and care must be taken not to break the adsorbent surface. Only two spots can be placed on a microscope slide (Figure 8*b*), and these should be the same distance from the lower edge. On larger plates, spots should be *ca.* 1 cm apart.

3. Developing the chromatogram. Large TLC plates are developed in special tanks and jars but a wide-necked screw-cap bottle (4 oz, 60 x 70 mm) serves well for microscope slides. The jar should contain a little developing solvent (about 5 mm depth) and be partially lined with a piece of filter paper dipping into the solvent. This ensures that the atmosphere of the jar is saturated with solvent vapour.

When the plate is placed in the jar, the lower edge of the adsorbent layer must be under the solvent but the spots of applied substance must be above the solvent level (Figure 8*c*). The jar is loosely capped during development.

The plate is removed from the jar before the solvent front reaches the top of the adsorbent layer and the position of the solvent front is immediately marked with a sharp-pointed instrument. The solvent is allowed to evaporate from the plate, preferably in a fume cupboard.

4. Location of components after development. The separation of coloured compounds is immediately apparent but special methods must be used to locate colourless substances.

(i) The TLC plate should first be viewed under an ultraviolet lamp when fluorescent compounds can be seen.

(ii) When the plate is allowed to stand in a developing jar containing crystals of iodine, the iodine vapour slowly dissolves in most organic compounds which thus show up as brown spots.

(iii) Many spray reagents have been described. Some are specific for compounds of a single type, others are more generally applicable: some interact reversibly with the compounds to be detected, others react destructively. Dichlorofluorescein, used in dilute ethanolic solution, is a valuable general non-destructive reagent. After spraying, the plate should be viewed under ultraviolet light.

Gas liquid chromatography (GLC)

Gas liquid chromatography has made great strides since its inception in the early 1950s and is invaluable for the qualitative and quantitative analysis of many compounds of suitable volatility. This means, in general, compounds of molecular weight up to several hundred and, in special cases, over 1000. Separation is based on partition of the components of a mixture between the liquid stationary phase and the gaseous mobile phase and is mainly dependent on the relative volatility of the components being separated.

The essential features of a gas chromatographic unit are shown in Figure 9. The column, consisting of a long glass or metal tube and usually coiled so that it can be

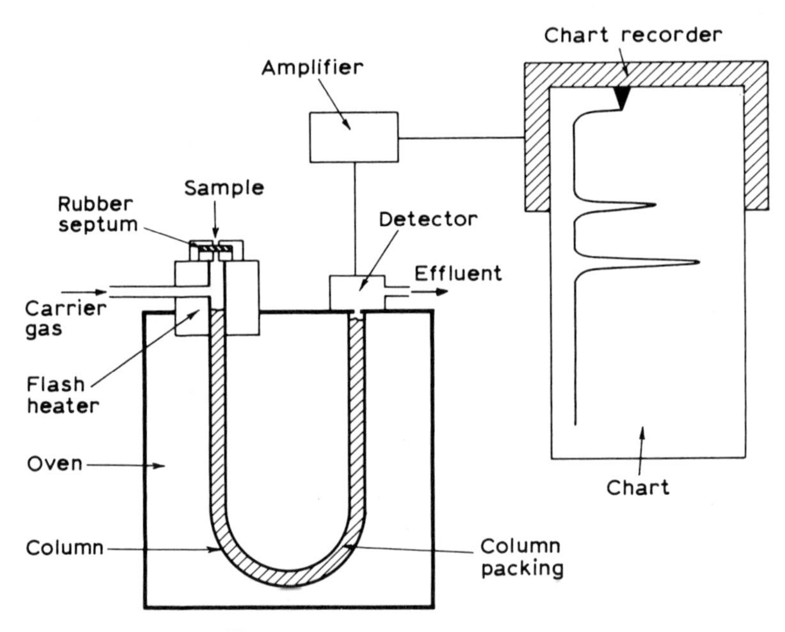

Figure 9. *Gas liquid chromatography*

fitted into a compact oven, is filled with column packing. This last is inert support material, such as some form of celite, coated with the stationary phase which is generally a viscous oil of low volatility. Silicone oils, Apiezon (hydrocarbon) greases, and polyesters of high molecular weight are commonly used for this purpose.

The column is mounted in an oven which can be accurately held at any temperature between room temperature and 250° (usually). The carrier gas (mobile phase) is usually nitrogen (but sometimes argon, helium, or hydrogen) and passes continuously through the column throughout its operation.

A small volume of the mixture to be separated is injected through a rubber septum into the column where it is vaporized, sometimes with the help of a flash heater. The components of the mixture partition themselves between the stationary and mobile phases and move through the column, ideally at different rates, separating into discrete bands. More volatile components are eluted before less volatile compounds, though secondary factors such as polarity may sometimes invert the expected order. The column effluent passes through the detector (Figure 9) which senses the separated components as they are eluted and sends an electric signal through an amplifier to a chart recorder. A typical chromatogram is shown in Figure 10.

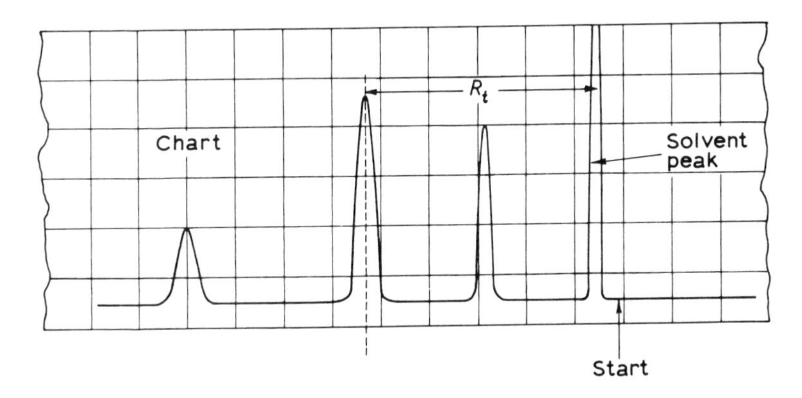

Figure 10. *Typical GLC trace*

1. Qualitative analysis. GLC is commonly used for qualitative analysis of mixtures. Every compound has a characteristic retention time (R_t in Figure 10) for a given set of operating conditions. This depends on:

(i) the nature and amount of stationary phase,
(ii) the flow rate of the carrier gas,
(iii) the temperature of the column.

These parameters should always be recorded on the chromatogram.

Two compounds with the same retention time under identical chromatographic conditions *may* be identical and this is more likely to be true if they retain this identity of behaviour on two different stationary phases (preferably a polar and a non-polar phase). Identity of chromatographic behaviour is conveniently observed by comparing the chromatogram of the mixture under study with that of the same mixture to which an authentic substance has been added. Identical substances give a single peak. Identity can hardly ever be proved by GLC alone; other physical and/or spectral properties should also be examined.

Material separated by GLC can be collected with special equipment. For this it is usual to inject larger quantities of material, and preparative apparatus is usually equipped with bigger columns and with the facility of repeated and automatic injection of the sample and automatic collection of the separated components.

2. Quantitative analysis. Lengthy discussion on the quantitative application of GLC is beyond the scope of this book. The area under each peak of a GLC trace is related to the amount of material being eluted but the precise relationship must generally be determined with standard mixtures. A simple and approximate technique which can be used for closely related compounds is employed in Experiment 6.

3. Some practical points. Since operating conditions vary with the instrument and detector being used, details are not given here. The following points deal with the selection of a column and its operating temperature.

Column packings usually contain 2-20% of stationary phase. Higher loadings of stationary phase lead to longer retention times at any given temperature so that lightly loaded columns are useful for compounds of high boiling point. Columns with 10% of stationary phase operated 30-50° below the boiling point of the sample are generally suitable. The choice of stationary phase and of optimum temperature is a question of experience. Retention times are shorter at higher temperatures, but each stationary phase has a maximum operating temperature above which it is degraded or 'bleeds' off the column into the detector.

The longer a column, the greater its resolving power and the longer the retention time. Columns up to 2 m long are suitable for most analytical purposes and for difficult separations other stationary phases should be examined before a longer column is used.

Polar compounds which are eluted very slowly or not at all are sometimes converted into less polar derivatives for gas chromatography, e.g. carboxylic acids into esters, and alcohols into acetates or trimethylsilyl ethers.

5. Infrared and ultraviolet spectroscopy*

The absorption of electromagnetic radiation by a molecule results in an increase in the molecular energy. Since the energy of a molecule can have certain quantized values only, it follows that a given molecule will absorb radiation selectively, absorption occurring where the radiation quanta possess the correct energy to raise the molecular energy from one permitted value to another. The energy of the radiation is proportional to the frequency (inversely proportional to the wavelength), and so any molecule may be expected to absorb radiation at specific frequencies.

Infrared spectra

1. Theory. Molecular vibrations (bending and stretching of bonds) in organic compounds occur with frequencies corresponding to the infrared region of the spectrum. When a molecule is subjected to infrared radiation of a frequency corresponding to one of its natural vibration frequencies, then radiation is absorbed and that particular vibration excited. Different types of bond vibrate at different frequencies, and so it is possible, by observing the infrared absorption frequencies of a compound, to derive information about the types of bond in the molecule, and hence about its functional groups.

Most of the common organic functional groups absorb in the wavelength range 2.5-15 μm (frequency 1.2-0.2 x 10^{14} cycles per second). The positions of absorption maxima are more commonly quoted, however, in terms of *wavenumbers* ($\bar{\nu}$) and expressed in cm^{-1}:

$$\bar{\nu} = \frac{1}{\text{wavelength}} = \frac{\text{frequency}}{\text{velocity of light}}$$

* Useful supporting film: Molecular Spectroscopy (Chemistry Education Material Study, U.S.A.).

2. Procedure. The infrared spectrum of a compound is obtained by placing a cell containing the compound in a beam of infrared light, passing the radiation which is not absorbed, through a prism or diffraction grating to split it up into its component wavelengths, and thence to an electrical recording device.

It follows that the cell which contains the sample must not itself absorb infrared radiation, and glass and quartz cells are therefore unsuitable for infrared spectral measurements. The most convenient material for infrared cells is sodium chloride (in the form of polished rock salt plates), and *samples for infrared spectroscopy must therefore be dry* in order to avoid 'fogging' of the salt faces which reduces the amount of radiation transmitted.

3. Sample preparation. (a) *Liquids* are most conveniently examined by placing one drop of liquid on a polished salt plate, placing another plate on top of the first, and inserting the resulting 'sandwich' into the beam.

Liquids may also be examined in dilute solution in an inert solvent such as carbon tetrachloride, chloroform, or carbon disulphide, but in elementary work this is rarely necessary.

(b) *Solids* cannot usually be examined in the crystalline state, since in this form they scatter appreciable proportions of the radiation. The spectra of solids may be obtained for dilute solutions, as with liquids. Alternatively the solid may be ground with an alkali halide (usually potassium bromide) and the mixture compressed at high pressure into a transparent disc. Usually, however, the spectra are obtained for *mulls*, i.e. suspensions of the finely ground solids in an inert liquid.

Ideally a mulling liquid should be chemically inert, and transparent throughout the infrared region, but in practice the liquids used have a few absorptions themselves. The most common liquid is *Nujol* (a mixture of alkanes) which contains only C–C and C–H bonds and hence has a comparatively simple spectrum (Figure 11), the principal absorptions of which are associated with C–H vibrations.

Since almost all organic compounds contain C—H bonds, the absorptions of the compound which coincide with those of Nujol are of interest only occasionally. In those cases, however, other mulling agents, e.g. *hexachlorobutadiene* ($CCl_2{=}CCl.CCl{=}CCl_2$) may be used in place of Nujol.

(c) *Preparation of mulls.* A small quantity (*ca.* 5 mg) of the solid is finely ground in an agate mortar. One drop of the mulling liquid is then added, and thoroughly mixed with the powdered solid, giving a smooth creamy paste. Obtaining a mixture of the correct consistency is a matter of practice. If the mull is too thick (it will appear dry), a large proportion of the radiation will be scattered rather than transmitted, and the spectrum will be poorly resolved. If the mull is too thin, the absorptions of the compound will be very weak.

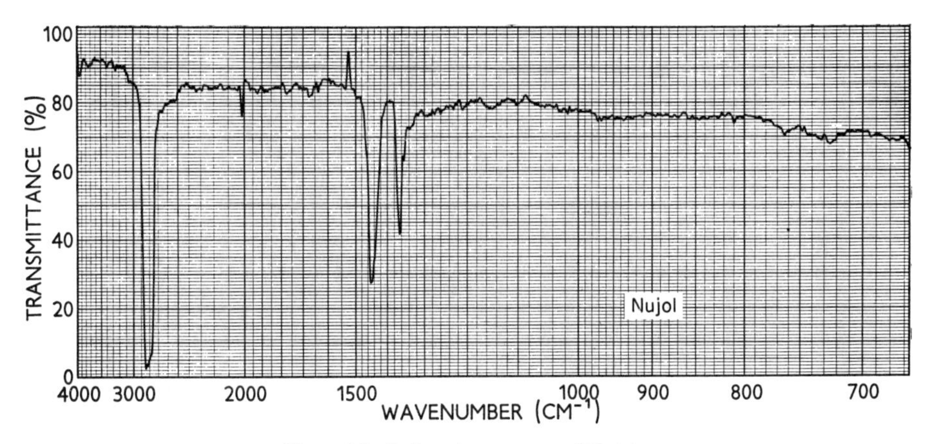

Figure 11. *Infrared spectrum of Nujol*

Interpretation of infrared spectra

A complex organic molecule, containing a large number of bonds of different types, will obviously be capable of vibrating in many ways, and hence will give rise to an infrared spectrum of some complexity. Not all the absorptions observed in the spectrum can be interpreted in terms of any particular vibrations, and many are of little diagnostic value in detecting functional groups. Where, however, the absorption excites the vibrations of a bond which joins a single atom or small group of atoms to the rest of the molecule, that absorption is significant in revealing such a bond in the molecule. The same is true of absorptions due to multiple (i.e. double and triple) bonds.

For the purposes of obtaining structural information from an infrared spectrum, it is convenient to divide the spectrum into five regions as shown in Figure 12. The principal absorptions observed in these regions are indicated on the chart.

Region 1: 4000-2850 cm^{-1} (2-3.5 μm). The principal absorptions in this region correspond to the stretching frequencies of *single bonds involving hydrogen:* C—H, O—H, and N—H.

C—H vibrations normally occur at the lower end of this range (below 3100 cm^{-1}). Since organic compounds almost invariably contain C—H bonds, any absorption observed in this region is of little value diagnostically (*although the complete absence*

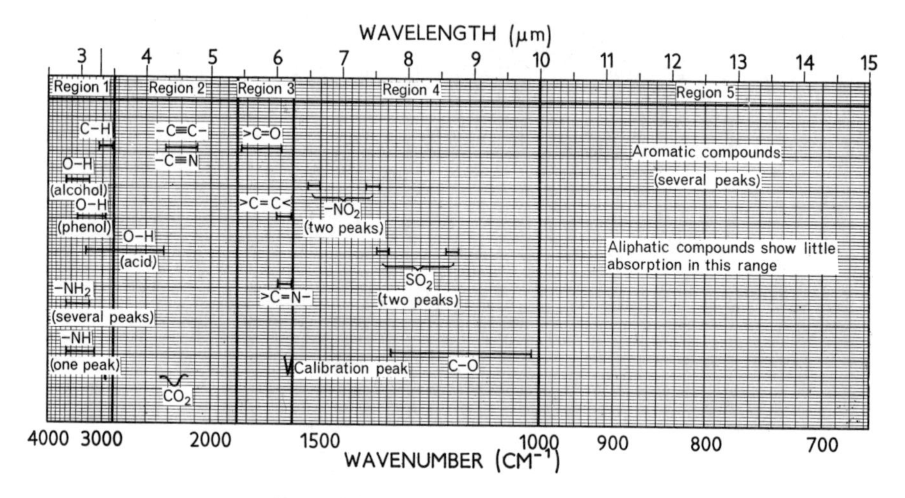

Figure 12. *Characteristic group frequencies*

of peaks in this range may be highly significant). It may be possible in some cases, however, to distinguish H—C (saturated) from H—C (unsaturated) bonds; the former absorb at *ca.* 2900 cm^{-1} (*cf.* Nujol, Figure 11) and the latter at *ca.* 3000-3100 cm^{-1} (cf. benzaldehyde, p. 39).

The only notable exceptions to this range are the C—H bonds of terminal acetylenes, RC≡CH, which absorb at 3300 cm^{-1} (and may therefore be confused with N—H), and aldehydes R—C(=O)H, which absorb between 2700 and 2800 cm^{-1} (cf. benzaldehyde, p. 39; but these absorptions are often weak and not easily recognized).

If the spectrum shows strong absorption above *ca.* 3050 cm^{-1}, it almost certainly contains an O—H or N—H or NH_2 group.

O—H vibrations fall conveniently into three classes, depending on the environment of the —OH group: (*a*) alcohols, (*b*) phenols, (*c*) carboxylic acids. As the acidity of the hydroxyl hydrogen increases (alcohol < phenol < acid) and the tendency to form intermolecular hydrogen bonds increases, so the frequency of the O—H absorption decreases and the peak becomes broader. Thus alcohols usually absorb at 3400-3200 cm^{-1}, phenols at 3200-3000 cm^{-1}, and carboxylic acids at 3000-2500 cm^{-1} (see Figure 13).

(Dilute solutions of alcohols and phenols, in which intermolecular hydrogen bonding is minimized, give O—H stretching absorption as a sharp peak in the region of 3600 cm^{-1}).

N—H vibrations in amines and amides normally absorb within the range 3500-3150 cm^{-1}. Hydrogen bonding affects the frequency of N—H absorption much less than O—H absorption. N—H peaks are usually sharper but weaker in intensity than O—H peaks.

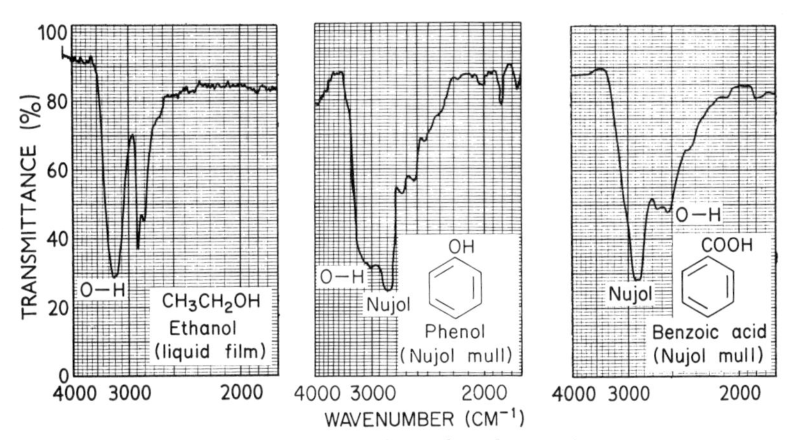

Figure 13. *O-H stretching frequencies*

It is possible to distinguish between primary and secondary amines and amides on the basis of these N—H absorptions. The —NH— group (secondary) gives rise to a single peak (e.g. *N*-methylaniline) and the $-NH_2$ group (primary) to a multiplet (e.g. aniline). In dilute solutions, where hydrogen bonding is at a minimum, $-NH_2$ absorptions appear as a sharp doublet, but otherwise the usual absorption pattern is as shown in Figure 14. It should be noted that tertiary amines and amides, which do not contain N—H bonds, do not absorb in this region.

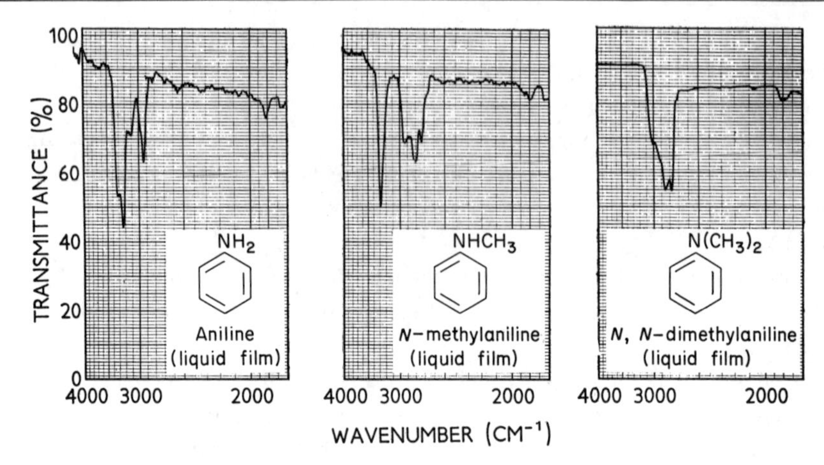

Figure 14. *Infrared spectra of amines*

Region 2: 2850-1850 cm^{-1}. (3.5-5.4 μm). Although strongly hydrogen-bonded O—H group absorption (e.g. in acids) extends down to *ca.* 2500 cm^{-1}, in general this is a region in which few molecules absorb at all. The only common functional groups which have characteristic absorptions within this region are *triple bonds* (C≡C and C≡N), which absorb between 2100 and 2300 cm^{-1}, and these are often of low intensity except those which form part of a conjugated system or those of terminal acetylenes.

Thiols absorb (S—H stretching) at 2500-2600 cm^{-1}; and *cumulated double bond systems* (X=C=Y) absorb in the range 1900-2300 cm^{-1}, but these groupings are rarely encountered in elementary identifications.

Some spectra may show a small, broad peak at *ca.* 2300 cm^{-1} (cf. Figure 15); this is caused by atmospheric carbon dioxide and is to be disregarded.

Region 3: 1850-1600 cm^{-1} (5.4-6.25 μm). This narrow range of frequencies is of great importance because it contains double-bond stretching vibrations, of which the most frequently encountered are C=C, C=N, and C=O.

By far the most useful of these for the purpose of detecting functional groups is *carbonyl group absorption.* This is almost invariably a very strong absorption and occurs in a region (1650-1850 cm^{-1}) where no other groups absorb with comparable intensity.

The exact frequency of carbonyl absorption is dependent on the molecular environment of the carbonyl group, and it is thus often possible to obtain, from the frequency, a clue to the type of carbonyl function present. The table gives an approximate frequency range for different types of carbonyl groups.

anhydrides	1850-1740 cm^{-1} (2 peaks, *ca.* 60 cm^{-1} apart)
acyl halides	1815-1750 cm^{-1} (1 peak)
esters	1750-1710 cm^{-1}
aldehydes	1740-1680 cm^{-1}
ketones	1725-1660 cm^{-1}
carboxylic acids	1720-1660 cm^{-1}
amides	1680-1630 cm^{-1}

The spectra of typical carbonyl compounds are shown in the collection of spectra on pp. 38-47 (see also Experiment 13).

Conjugation of the carbonyl group with a π-electron system lowers the carbonyl absorption frequency; $\alpha\beta$-unsaturated and aromatic carbonyls thus absorb towards the lower end of each frequency range. Hydrogen bonding (especially intramolecular) also lowers the frequency, in some cases appreciably (cf. ethyl *o*-aminobenzoate, Figure 17, p. 35). On the other hand, incorporation of the carbonyl group into a strained ring system increases the frequency; thus cyclohexanone (p. 41) absorbs at the frequency (1710 cm^{-1}) expected for an acyclic ketone, whereas cyclopentanone absorbs at 1740 cm^{-1}, cyclobutanone at 1780 cm^{-1}, and cyclopropanone at *ca.* 1810 cm^{-1}.

C=C and C=N groups normally absorb at lower frequencies than C=O groups. They usually fall within the range 1690-1600 cm^{-1} and most often below 1670 cm^{-1}. These absorptions are frequently weak, and less intense than those of the C=O groups, are variable in position, and are generally of less value than C=O absorptions for the detection of the respective functional groups.

Some (although not all) aromatic compounds show a very sharp absorption at *ca.* 1600 cm^{-1}. This absorption in polystyrene (1603 cm^{-1}) is often used to calibrate the spectrophotometer chart.

Care must be exercised in interpreting absorptions in the 1650-1600 cm^{-1} range if N–H absorption has been observed in region 1, since the bending vibration of N–H bonds falls within this range.

Region 4: 1600-1000 cm^{-1} (6.25-10 μm). Both this region and region 5 often contain a complicated absorption pattern, only a few of these absorptions being of diagnostic significance. Outstanding among these are the absorptions of the *nitro-group* (1560-1500 *and* 1360-1320 cm^{-1}: two peaks) and the *sulphonyl* group (1350-1300 *and* 1160-1140 cm^{-1}: two peaks), and further sharp aromatic absorption, in some compounds, at *ca.* 1500 cm^{-1}. C—H bending (cf. Nujol, p. 27) and C—O and C—N stretching vibrations fall within this region but are not generally useful for identification purposes.

Region 5: 1000-667 cm^{-1} (10-15 μm). The most significant absorptions here are those of (unsaturated C)—H bending and C-halogen (especially C—Cl) stretching. Thus aliphatic compounds which contain neither multiple bonds nor halogens exhibit little absorption in the region; on the other hand, most aromatic compounds absorb strongly. It may often be possible to recognize the characteristic absorptions of a particular type of $=\overset{|}{C}-H$; some typical values are given below. It is usually sufficient, however, for elementary identifications, to use this region to distinguish qualitatively between aromatic and aliphatic compounds rather than to attempt a detailed analysis of the absorption pattern.

alkenes

$RCH{=}CH_2$ *ca.* 990 *and* 910 cm^{-1}

$RCH{=}CHR$ (*cis*) *ca.* 690 cm^{-1}
$RCH{=}CHR$ (*trans*) 980-960 cm^{-1}
$R_2C{=}CH_2$ *ca.* 890 cm^{-1}
$R_2C{=}CHR$ 840-810 cm^{-1}

aromatics

5 adjacent hydrogen atoms (i.e. monosubstituted)	770-730 and *ca.* 700 cm^{-1}
4 adjacent hydrogen atoms (i.e. *ortho*-disubstituted)	770-735 cm^{-1}
3 adjacent hydrogen atoms (i.e. *meta*-disubstituted and 1,2,3-trisubstituted)	810-750 cm^{-1}
2 adjacent hydrogen atoms (*para*-disubstituted, 1,2,4-trisubstituted and 1,2,3,4-tetrasubstituted)	860-800 cm^{-1}
Isolated hydrogen atom between two other substituents	*ca.* 880 cm^{-1}

Carbon-chlorine vibrations occur in the 800-600 cm^{-1} region but these are of no value for detecting chlorine in the molecule; this is much better done by other methods (see p. 145).

The spectrum as a whole. However complicated the absorption pattern of a particular compound, it is completely characteristic of that compound. The region of the spectrum below 1500 cm^{-1}, which is often very complex, is referred to as the *fingerprint* region; although two compounds containing the same functional groups may show very similar absorptions above 1500 cm^{-1}, even small differences in the structures of the compounds will give rise to different absorptions in the fingerprint region.

Confirmation of the identity of an unknown compound may thus be obtained by comparison of the infrared spectrum of the unknown with that of an authentic sample.

Examples of interpretation

The following examples show the kind of information which may be obtained from infrared spectra.

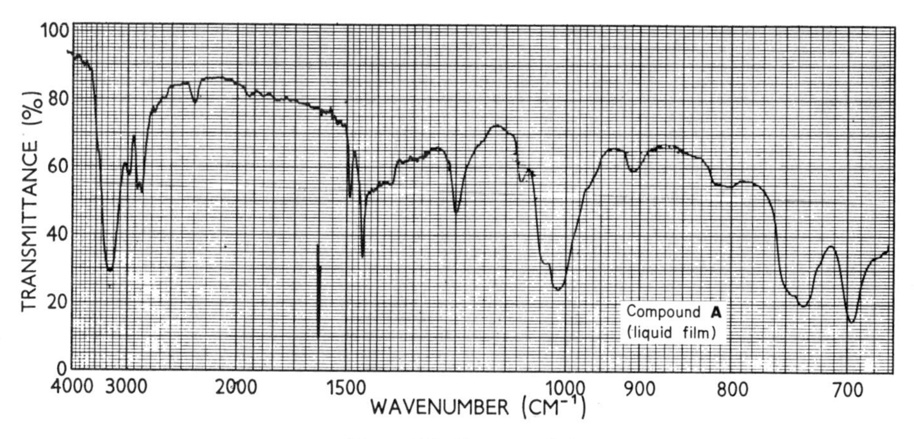

Figure 15. *Compound A*

Compound A (Figure 15). Neutral colourless liquid; nitrogen, sulphur, and halogens absent.

The comparative simplicity of the spectrum indicates that the molecule contains comparatively few types of bond. The strong broad absorption at 3250 cm^{-1} (region

1) is indicative of an alcoholic hydroxyl group, and the lack of absorption in region 3 indicates the absence of a carbonyl group. The large peak at 1010 cm^{-1} is probably the C—O stretching frequency associated with the hydroxyl function, and the two intense absorptions at 695 and 735 cm^{-1} suggest an aromatic system, probably a phenyl group (cf. p. 32, and the spectra of monosubstituted aromatic compounds on pp. 38-47).

The evidence thus suggests that this compound is an alcohol containing an aromatic (probably phenyl) substituent. (A is benzyl alcohol).

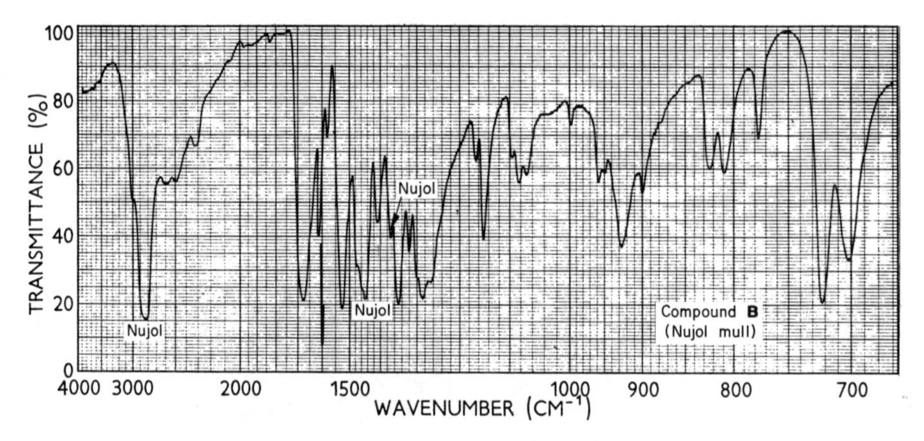

Figure 16. *Compound B*

Compound B (Figure 16). (Nujol mull). Almost colourless solid. Acidic (gives carbon dioxide with sodium bicarbonate). Nitrogen present; sulphur and halogens absent.

The observed acidity is suggestive of a carboxyl group, and the spectrum confirms this (broad O—H peak at <3000 cm^{-1}, and C=O absorption at 1675 cm^{-1}). The absorptions between 700 and 750 cm^{-1} indicate an aromatic compound. The nitrogen is not present as a primary or secondary amino-group (no N—H absorption) but the strong peaks at 1525 and 1350 cm^{-1} suggest the presence of a nitro-group.

Compound B is thus probably an aromatic carboxylic acid with a nitro-substituent. (B is *m*-nitrobenzoic acid.)

Compound C (Figure 17). Basic colourless liquid; nitrogen present.

The basicity and the presence of nitrogen suggest that C is an amine. Region 1 of

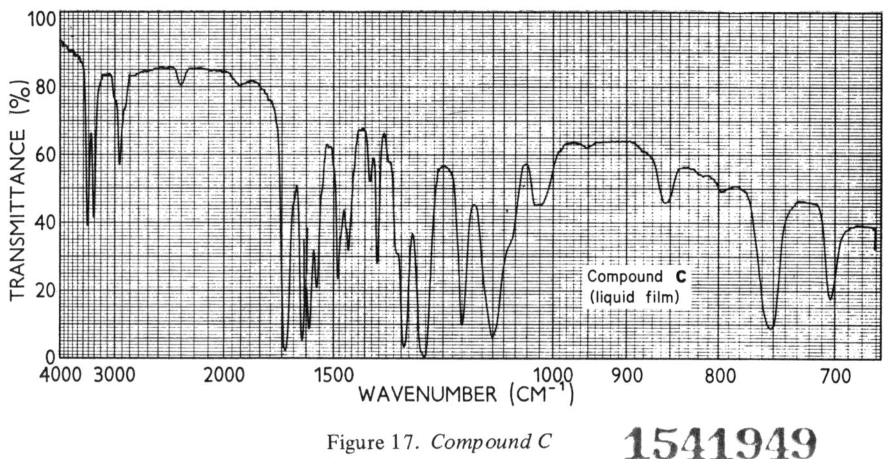

Figure 17. *Compound C*

the spectrum indicates that the amine is primary, and region 3 indicates that C also contains a carbonyl group ($\bar{\nu}_{max}$ 1680 cm^{-1}). It is difficult to deduce the exact nature of this carbonyl group from the spectrum alone; chemical tests are necessary to establish this. The compound apparently contains an aromatic system ($\bar{\nu}_{max}$ 755 and 705 cm^{-1}); the spectrum does not indicate whether the amino- or carbonyl functions are substituents on the aromatic ring(s) or on aliphatic side chains.

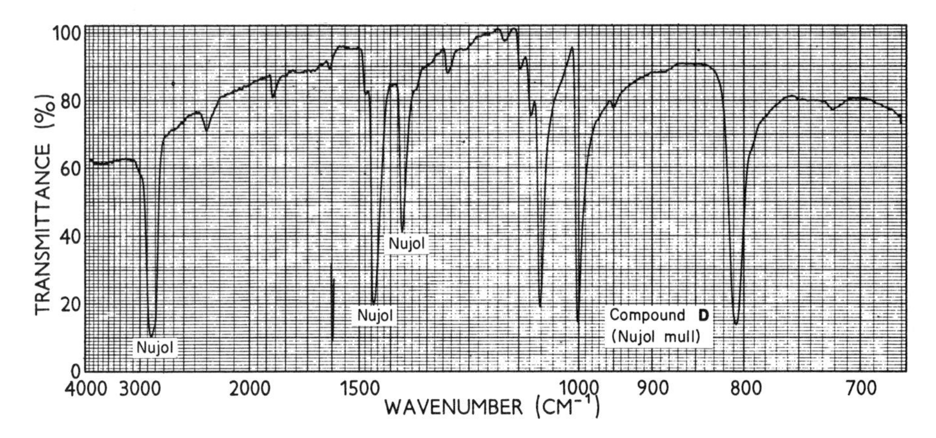

Figure 18. *Compound D*

Compound C is thus an aromatic compound containing NH_2 and C=O groups. (C is ethyl *o*-aminobenzoate. Why does $\bar{\nu}_{max}$ (C=O) occur at such low frequency and why is the NH_2 absorption a simple doublet? See p. 31 or consult your instructor.)

Compound D (Figure 18). (Nujol mull). Neutral colourless solid: bromine present.

It is difficult to tell from this spectrum what functional groups are present; it is, however, comparatively easy to tell which groups are absent. Absence of absorption (Nujol and other C—H vibrations excepted) above 1100 cm^{-1} rules out the presence of most functional groups. The only significant peak is that at 810 cm^{-1}, characteristic of an aromatic compound with *at most* three adjacent hydrogens.

Compound D thus appears to be a simple aromatic bromo-compound. (D is *p*-dibromobenzene).

Ultraviolet spectra

Whereas the absorption of infrared radiation by molecules results in excitation of molecular vibrations (see p. 25), the absorption of ultraviolet (and visible) radiation leads to the excitation of electrons from their normal energy levels (ground state) to higher-energy states. The electrons in these higher energy states occupy *antibonding orbitals*; these are unoccupied in the ground state. Normally the electrons forming single (σ) bonds require radiation of very high energy (i.e. short wavelength, <150 nm) for promotion to an excited (σ^*) state, and so compounds containing only σ-bonded electrons exhibit no absorption in the normal ultraviolet/visible region (200-800 nm).

Excitations of π-electrons and non-bonding (n) electrons are transitions requiring lower energy, and are thus effected by radiation in or near the ordinary ultraviolet region. For example ethylene absorbs at 171 nm ($\pi \rightarrow \pi^*$ excitation) and trimethylamine at 227 nm ($n \rightarrow \sigma^*$ excitation). Molecules such as carbonyl compounds, in which an atom involved in π-bonding also possesses a lone pair of electrons, can exhibit $n \rightarrow \pi^*$ absorption, in addition to $n \rightarrow \sigma^*$ and $\pi \rightarrow \pi^*$ absorption.

Where π-electrons are delocalized (i.e. where the moleule possesses a system of conjugated multiple bonds), the energy levels of the bonding and antibonding orbitals are closer together, the energy required for electronic excitation is further diminished, and so conjugated compounds exhibit appreciable absorption in the ultraviolet region. In general the greater the degree of delocalization, the longer is the wavelength of the absorption maximum; thus compounds containing extended systems of conjugation may absorb, at least to a certain extent, at wavelengths above 400 nm, i.e. in the visible region of the spectrum, and hence appear coloured to the eye.

The chart obtained from most recording ultraviolet/visible spectrophotometers gives a plot of *absorbance* or *optical density* versus wavelength. The absorbance (A) is dependent on the concentration (c) of the absorbing substance in the cell, and also on the thickness (l) of the cell: thus

$$A = \epsilon c l.$$

The proportionality constant, ϵ, is called the *extinction coefficient* and is a measure of the intensity of absorption at a particular wavelength. It is usual to calculate *molar extinction coefficients*, by expressing concentration in moles per litre, and l in cm.

The ultraviolet spectra of organic compounds are almost always recorded for dilute solutions. Common solvents for ultraviolet spectroscopy are simple alcohols and saturated hydrocarbons. It is usual to use silica cells, since glass is not transparent throughout the ultraviolet region.

An exercise in the determination of ultraviolet spectra is included in Experiment 47.

Examples of infrared spectra

In addition to the following spectra, which appear in alphabetical order, attention is drawn to the spectrum of Nujol (Figure 11) on p. 27.

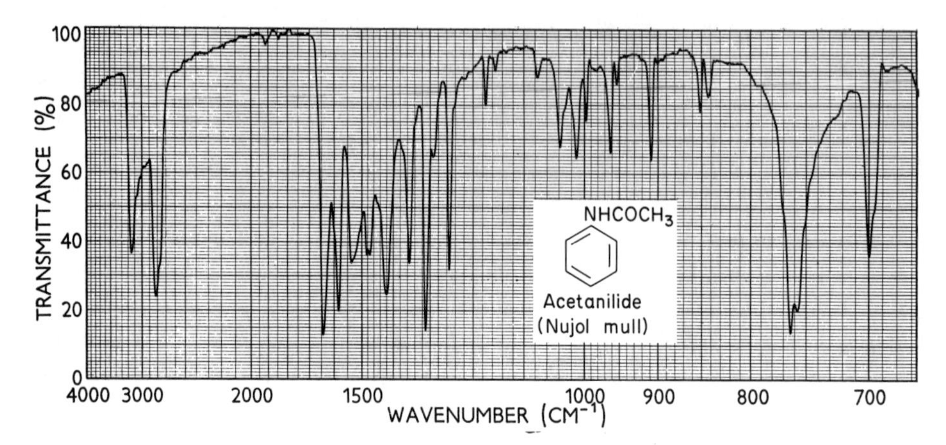

Figure 19. *Acetanilide*

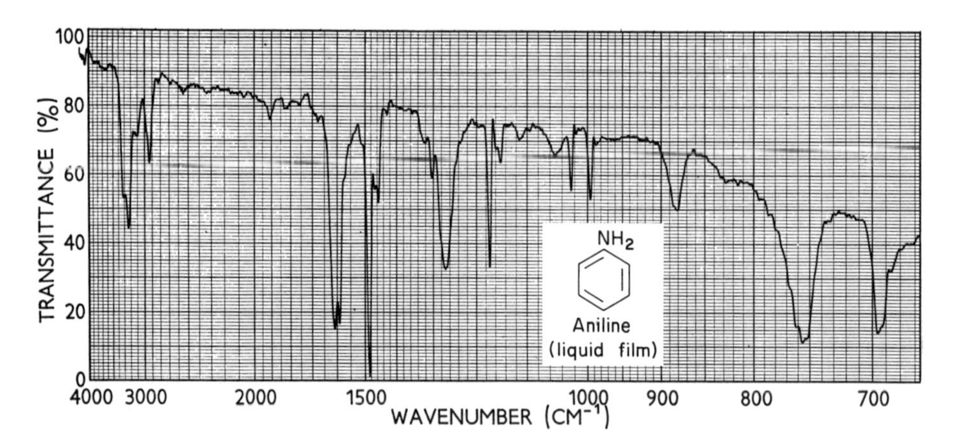

Figure 20. *Aniline*

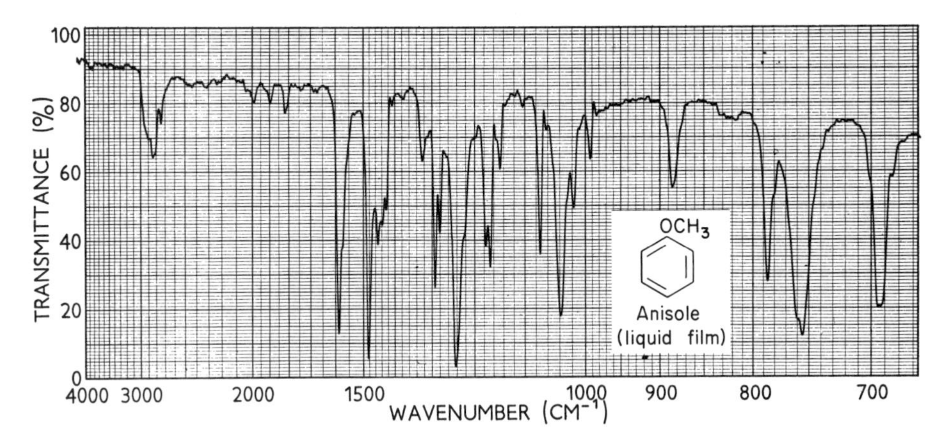

Figure 21. *Anisole*

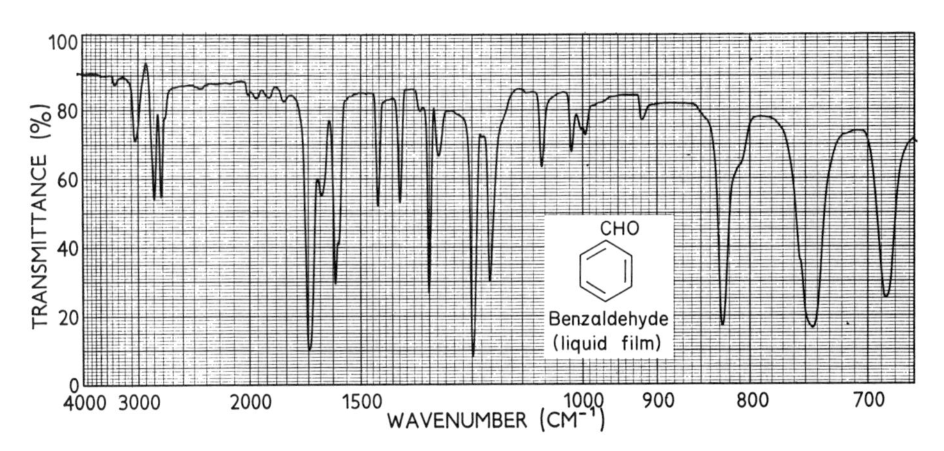

Figure 22. *Benzaldehyde*

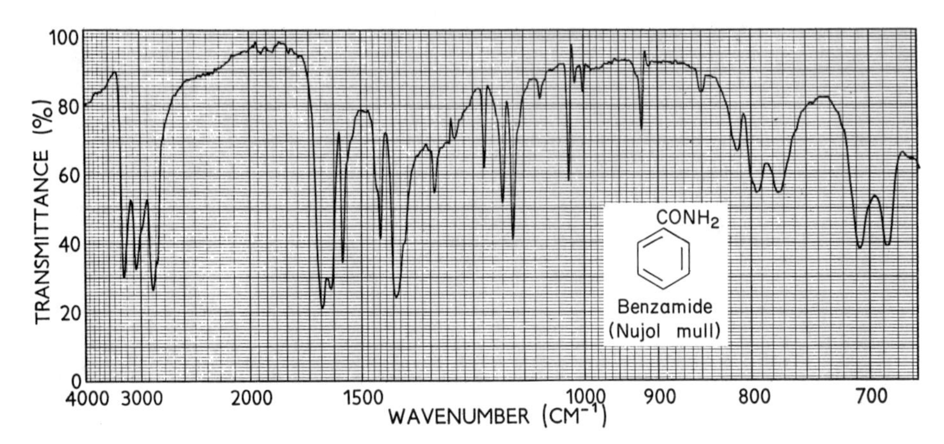

Figure 23. *Benzamide*

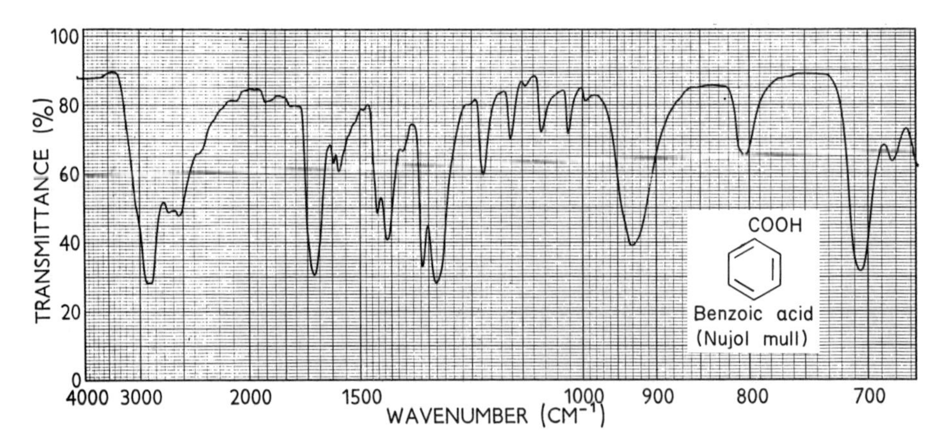

Figure 24. *Benzoic acid*

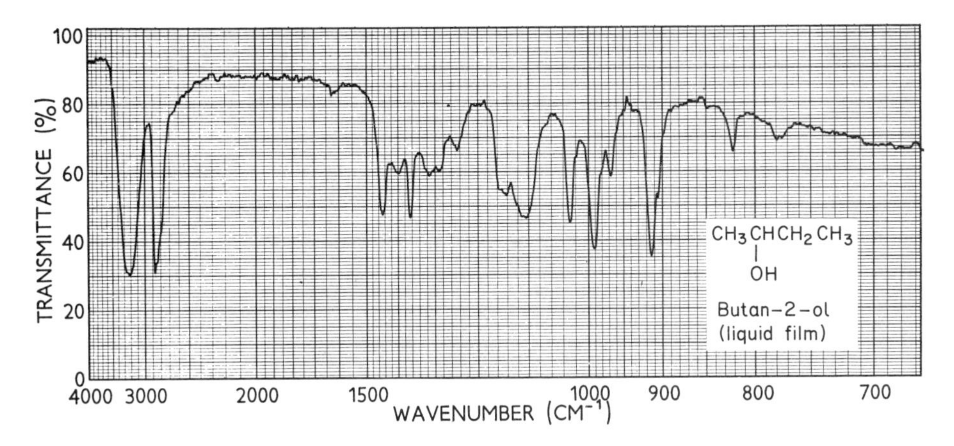

Figure 25. *Butan-2-ol*

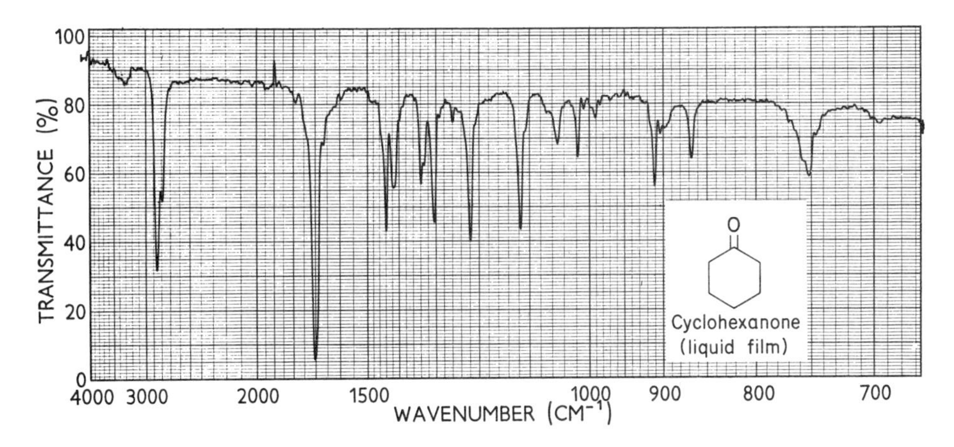

Figure 26. *Cyclohexanone*

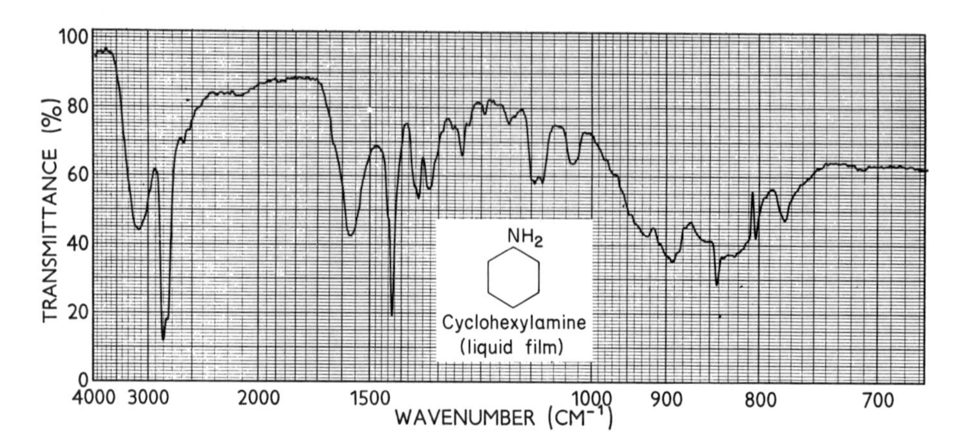

Figure 27. *Cyclohexylamine*

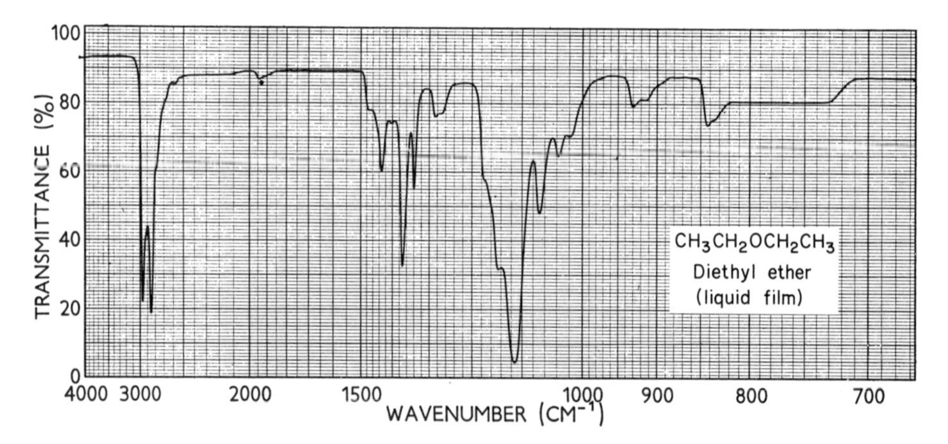

Figure 28. *Diethyl ether*

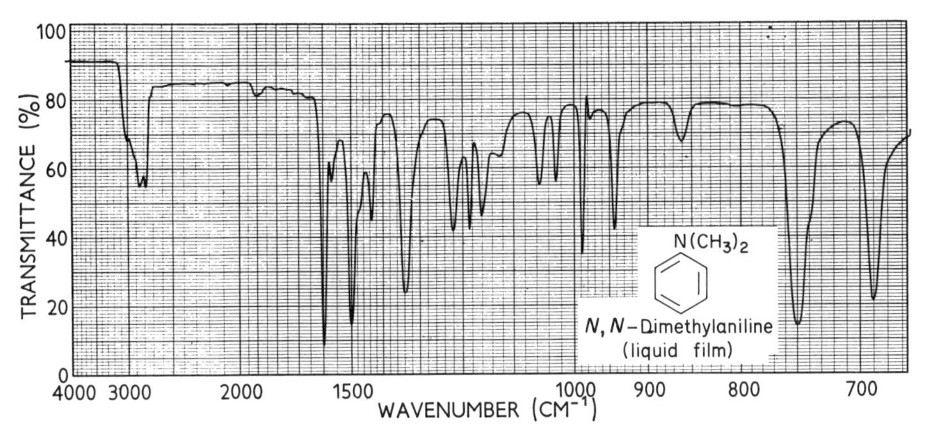

Figure 29. N,N-*Dimethylaniline*

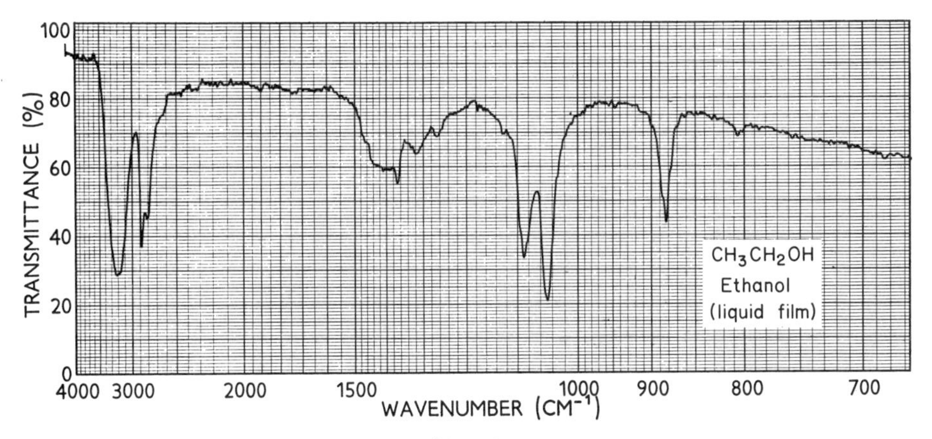

Figure 30. *Ethanol*

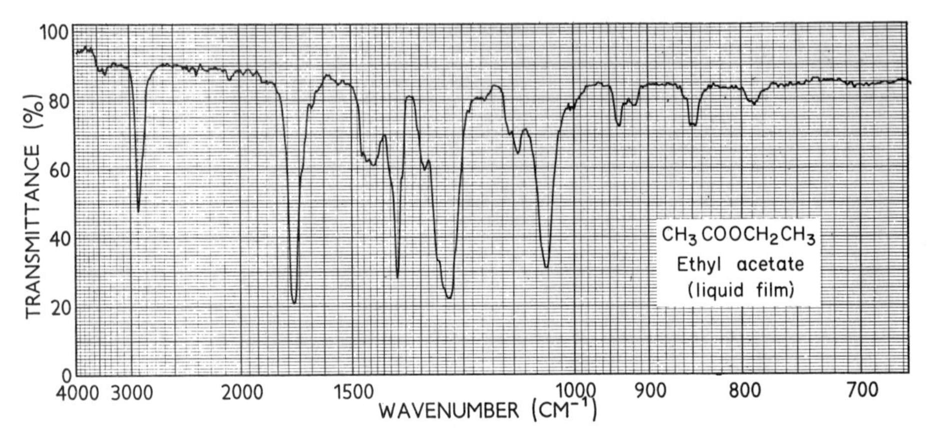

Figure 31. *Ethyl acetate*

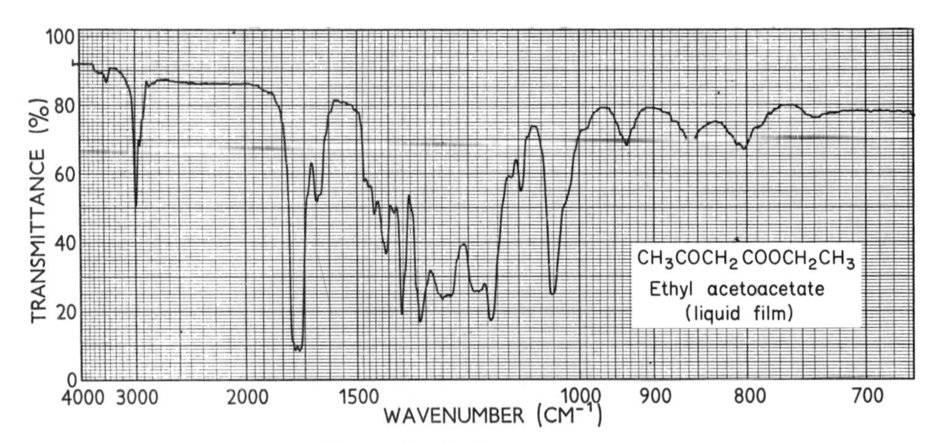

Figure 32. *Ethyl acetoacetate*

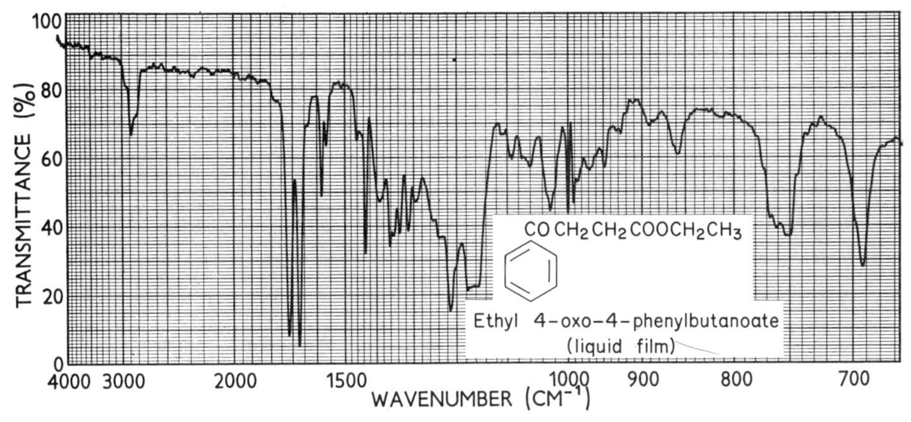

Figure 33. *Ethyl 4-oxo-4-phenylbutanoate*

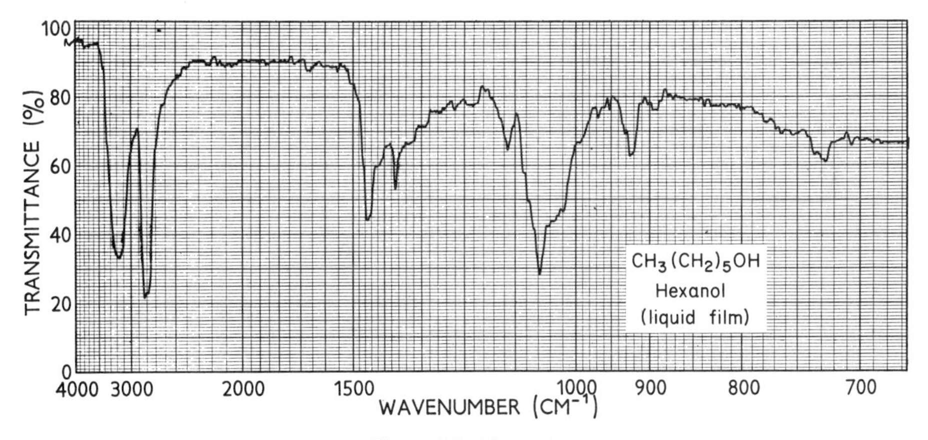

Figure 34. *Hexanol*

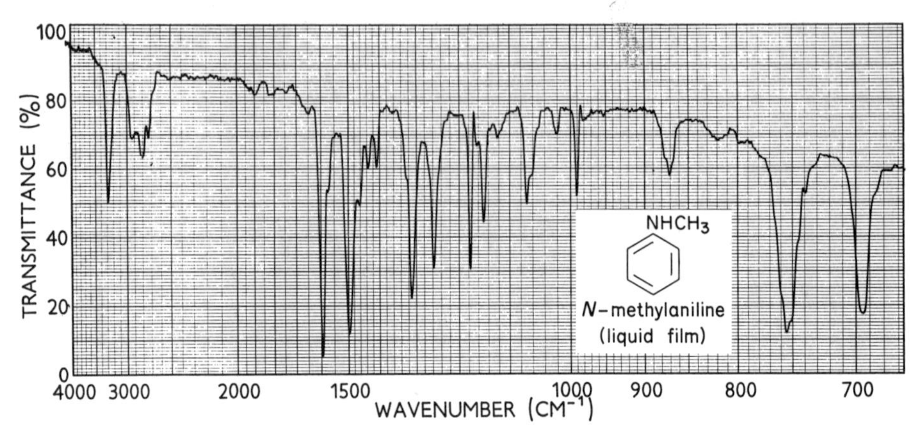

Figure 35. N-*Methylaniline*

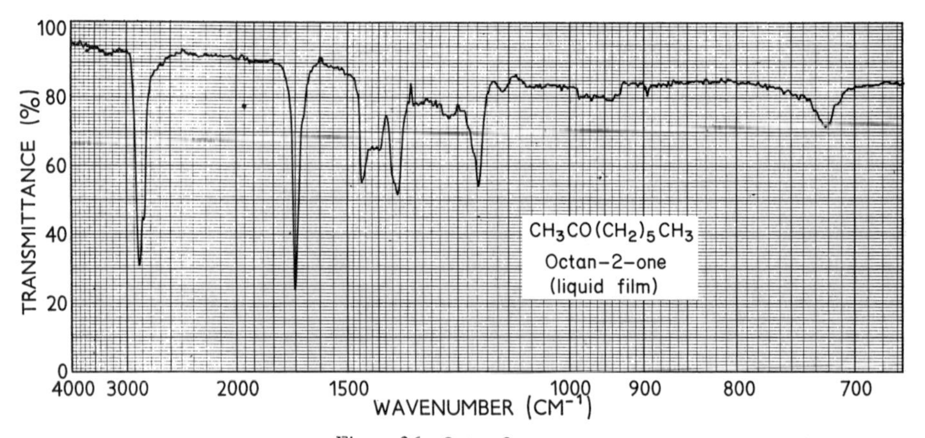

Figure 36. *Octan-2-one*

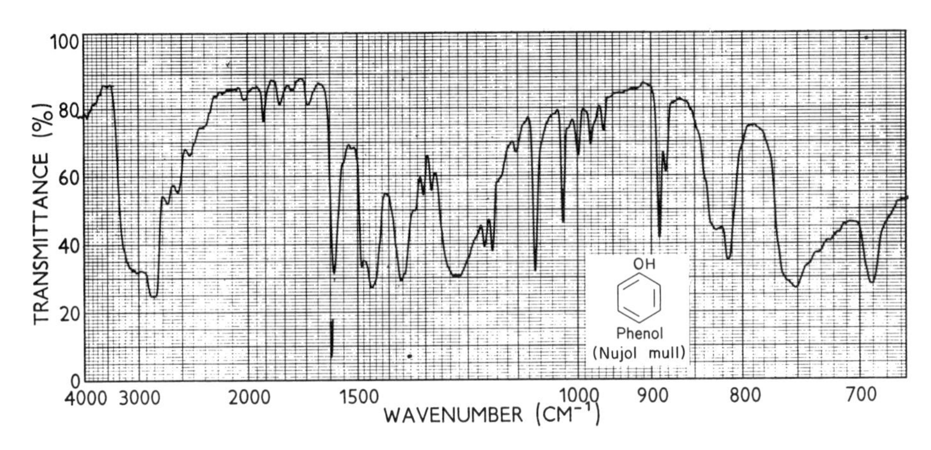

Figure 37. *Phenol*

6. Nuclear magnetic resonance spectroscopy*

Although the use of nuclear magnetic resonance (n.m.r.) spectroscopy in organic chemistry was introduced comparatively recently (in the 1950s), the value of the technique in providing structural information has led to its extensive use in the identification of organic compounds. Although the nuclei of several isotopes give rise to n.m.r. spectra, hydrogen is the only one of these nuclei which is commonly encountered in elementary organic chemistry, and so this discussion will be restricted to proton magnetic resonance (p.m.r.) spectroscopy.

The hydrogen nucleus may be considered as a spinning charged body, and as such has a magnet moment μ. It may be regarded as a tiny bar magnet. Its spin quantum number, I, is $\frac{1}{2}$, which means that when it is placed in a magnetic field it can take up only two orientations, i.e. lined up either with or against the field. The difference in energy between the two orientations, ΔE, is $2\mu H$, where H is the external field strength, and the absorption of energy by the nucleus to effect transition from one orientation to the other gives rise to the p.m.r. spectrum. (The Boltzmann distribution between the two possible energy states means that there is always a very small excess of nuclei lined up with the field, i.e. in the lower energy state, and hence there is a net *absorption* rather than a net *emission* of energy.) In practice, transitions are induced by subjecting the nucleus to a rotating magnetic field at right angles to the main magnetic field H, the frequency of the rotation, ν, being given by $\nu = \Delta E/h$, where h is Planck's constant.†

Thus $$\nu = \frac{2\mu H}{h} \qquad (1)$$

* Useful supporting film: Nuclear Magnetic Resonance (John Wiley & Sons Ltd.).

† For further information the reader is advised to consult a text-book on n.m.r. spectroscopy, e.g. J. R. Dyer, *Applications of Absorption Spectroscopy of Organic Compounds* (Prentice-Hall, 1965), Chapter 4, or L. M. Jackman and S. Sternhell, *Applications of Nuclear Magnetic Resonance Spectroscopy in Organic Chemistry* (Pergamon Press, Second Edition, 1969), Chapter 1.

When this equation is satisfied, the nuclear magnet is said to be *in resonance* with the rotating magnetic field and is capable of absorbing energy from the latter.

Proton magnetic resonance spectra may be obtained either by keeping the frequency, ν, constant and varying the external field, H, or by keeping H constant and varying ν. Instruments which operate with $\nu = 60$ MHz* are commonly used; these require field strengths of *ca.* 14,000 gauss in order that equation (1) should be satisfied and the spectrum observed.

These spectra are of value to the chemist because, at a given frequency ν, the field strength required for proton resonance depends on the molecular environment of the proton. For example, in the spectrum of methyl acetate (Figure 38), the two types of hydrogen (a) and (b) absorb energy at different field strengths and so give two separate signals on the p.m.r. spectrum. This difference is due to the fact that protons in different molecular environments 'see' slightly different magnetic fields, i.e. the field

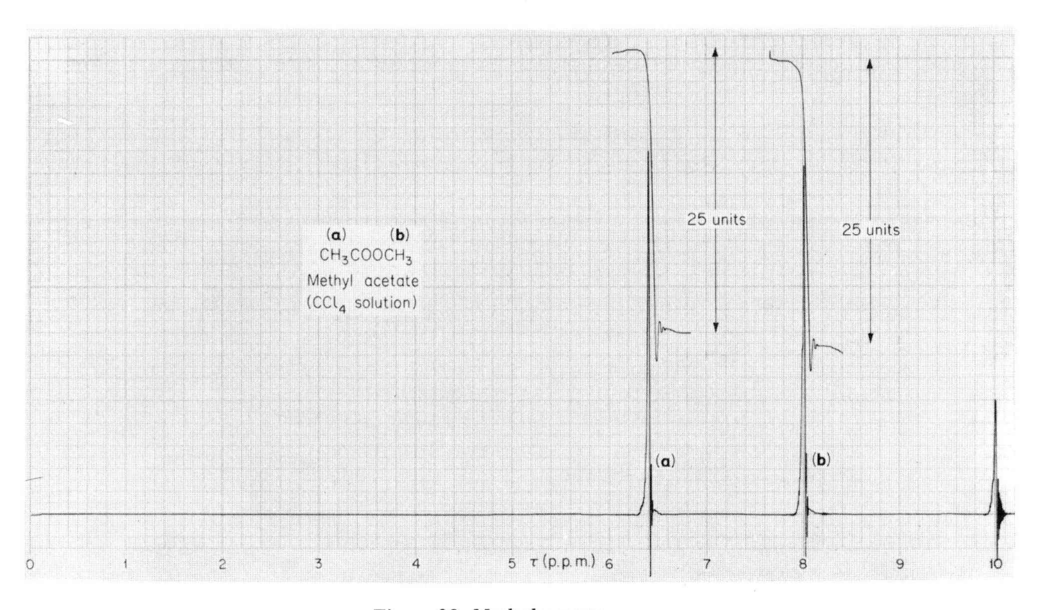

Figure 38. *Methyl acetate*

* 1 MHz = 1 million cycles per second.

experienced by a proton is slightly changed by the magnetic field within the molecule generated by movements of the electrons. This phenomenon is generally referred to as the 'shielding' effect of the electrons and arises because the external magnetic field induces a circulation of the electrons producing a small magnetic field of magnitude directly proportional to the external field and opposed to it.

N.m.r. spectra are normally obtained for dilute solutions. The solvents most commonly used are those which do not contain protons, e.g. carbon tetrachloride or deuteriochloroform ($CDCl_3$).

The positions of the peaks in the spectrum are usually measured from a reference peak and these separations are known as chemical shifts. The usual reference standard is tetramethylsilane, $SiMe_4$ (T.M.S.) which gives a single peak at a higher field strength than most organic compounds. Even although the spectrum may be scanned by keeping the frequency (ν) constant and changing the magnetic field (H) it is customary to calibrate p.m.r. spectra along the abscissa in frequency units (Hz); this is possible because the resonance frequency is proportional to field strength (equation (1)). However, since electronic shielding is proportional to the field strength (H), the chemical shift (in Hz) measured from T.M.S. will also be proportional to field strength and so will vary from one spectrometer to another. It greatly facilitates comparisons between spectra obtained on different instruments if the chemical shift values (Hz) are converted into dimensionless, field-independent units. This is done by dividing the chemical shift in Hz by the oscillator frequency of the spectrometer and expressing the result as p.p.m. (parts per million). It is conventional to assign the T.M.S. signal an arbitrary value of 10 p.p.m. and to express the chemical shift as a τ value where

$$\tau = 10 - \frac{\text{chemical shift in Hz from T.M.S.} \times 10^6}{\text{spectrometer oscillator frequency (in MHz)}}$$

[Another convention expresses the chemical shift (in p.p.m.) directly: it is denoted by the symbol δ. Thus $\tau = 10 - \delta$.]
Most protons in organic compounds absorb in the range 0 to 10τ but a few very acidic protons have negative τ valves.

The intensity of each signal (i.e. the area under the peak(s)) is proportional to the number of protons giving rise to that signal. Most spectrometers incorporate an *integrator* which measures the area under each peak and produces on the chart a 'stepped' trace where the height of each step is proportional to the number of protons giving that signal. For example in the spectrum of methyl acetate (Figure 38) the stepped trace (at the top of the spectrum) shows two steps of equal height (25 units), indicating that the two signals are produced by equal numbers of protons.

Figure 39 indicates typical τ values for some common functional groups, but these values must be used with some caution since in a complex molecule the observed τ values may differ appreciably from the values quoted. In particular, protons attached to carbons between two deshielding functional groups will be subject to the combined

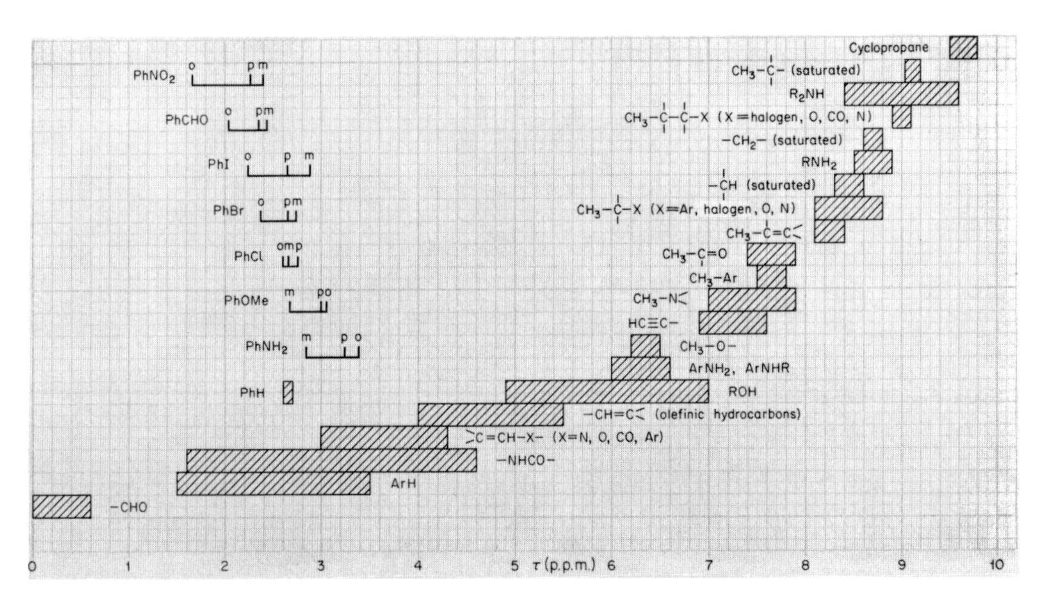

Figure 39. *Typical chemical shifts*

Acidic protons may have negative τ values, e.g.

R·COOH	0 to -3τ
R·C(–OH)=C·R′ (enols)	-5 to -9τ

Phenols (if intramolecularly hydrogen bonded) 0 to -5τ, otherwise 2 to 6τ.

effects of the neighbouring groups; for example, the methylene protons of diethyl malonate, $EtO_2CCH_2CO_2Et$, absorb at a much lower τ value (6.63τ) than the methylene protons of ethyl propionate, $CH_3CH_2\,CO_2Et$ (7.72τ).

Spin-spin coupling

The spectrum of ethyl acetate (Figure 40) is much more complex than that of methyl acetate (Figure 38), in that it has 8 lines appearing in three groups (4, 1, 3) rather than

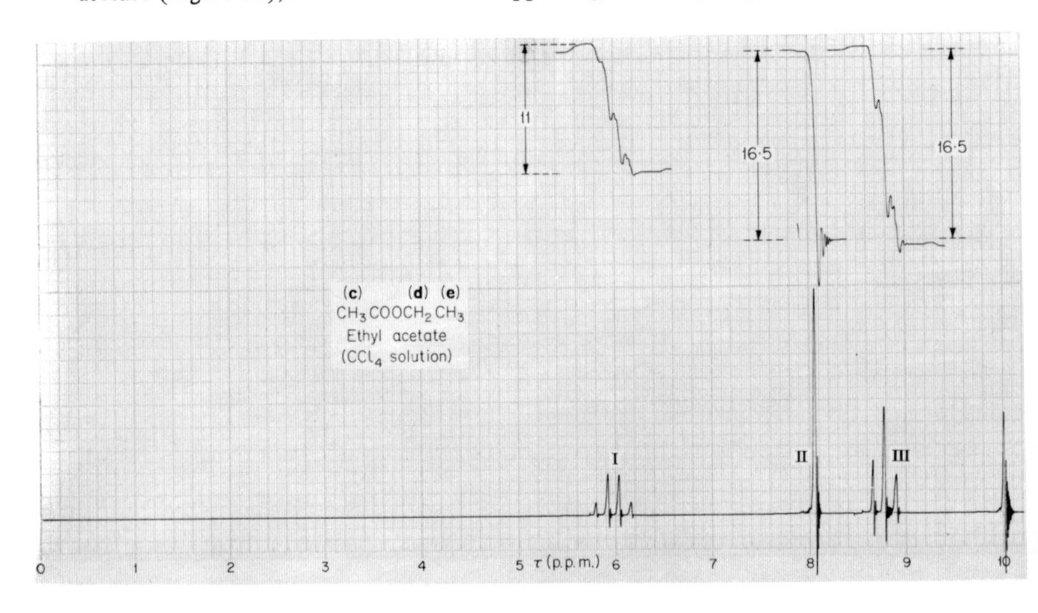

Figure 40. *Ethyl acetate*

the 3 lines which might be expected from the previous discussion. From Figure 39 and by comparison of Figures 38 and 40, it can readily be deduced that the (c) protons correspond to peak II, and (d) protons to group I and (e) protons to group III. This splitting of the ethyl group absorptions is due to a magnetic interaction between the protons on the adjacent carbon atoms.

Consider first the simple case of spin-spin coupling in the system $\overset{(f)}{CH}-\overset{(g)}{CH_2}-$, for example in Cl_2CH-CH_2Cl. The magnetic field experienced by the (g) protons will be affected by the orientation of the (f) proton. Since, at any instant, about half the molecules in the sample will have the (f) proton oriented with the external field (↑) and the remainder will have the (f) proton oriented against the field (↓), the absorption of the (g) protons appears as two equal lines, separated by about 7 Hz (this separation is known as the **coupling constant** J and is independent of the strength of the external field).

The chemical shift of the (g) protons is the mean of the τ values of the two peaks. Similarly there are three ways in which the two (g) protons can be oriented:

↑↑	both aligned with the field
↓↑ ↑↓	one with, one against the field
↓↓	both against the field

and hence the (f) absorption appears as a (1:2:1) triplet again with a coupling constant of 7 Hz; the chemical shift of the (f) proton is the mean τ value (symmetry indicates that this will be the τ value of the central peak). In a similar way in Figure 40 the (e) protons give rise to a triplet and the (d) protons to a quartet (1:3:3:1).

These ratios are observed only if the (f) and (g) protons have widely different chemical shifts (difference $\Delta\nu \geqslant 10\, J$). Where the chemical shifts are close, distortion of the signals takes place (even in ethylbenzene, Figure 45, where the difference is slightly greater than 10 J, the methyl group signal consists of a 2:3:1 triplet). As $\Delta\nu$ decreases, the distortion of each multiplet increases, and much more complex spectra result.

A fuller discussion of the interpretation of spin-spin splitting patterns is beyond the scope of this section but the following generalizations will be found useful:

(1) Protons in the same chemical and magnetic environment do not normally couple with one another.

(2) In saturated acyclic systems, coupling is important only between protons on the same and on adjacent atoms, but longer range coupling can be observed in unsaturated and other more rigid systems.

(3) For spectra where the differences in the chemical shifts of the nuclei are large compared with the coupling constants, the multiplicity of a group equals $(n + 1)$ where n is the number of protons on adjacent atoms.

(4) Protons undergoing rapid *chemical exchange* do not normally couple with adjacent protons. Chemical exchange is most commonly encountered in the

spectra of alcohols: in a given period of time any one hydroxyl proton may be attached to (and detached from) a number of alcohol molecules, especially if traces of acid or base are present. So the protons on the adjacent carbon are subjected to the effect of a rapid succession of hydroxyl protons, some of which have spins which are oriented with the field and others against it, and the net effect is that no splitting of the CH signal by the —OH proton and of the —OH signal by the CH proton(s) is observed.

A list of coupling constants is given in the following table.

Spin-spin coupling constants

		$J_{HH'}$ (Hz)
geminal coupling		
$>C<^{H'}_{H}$		10-15
$=C<^{H}_{H'}$		0-3
vicinal coupling		
$>$CH—CH'$<$	(free rotation about C–C bond)	5-8
=CH—CH'$<$	(free rotation about C–C bond)	4-10
=CH—CH'=		6-13
$>$CH—CH'(=O)		0-3
=CH—CH'(=O)		5-8

vicinal coupling–cont.

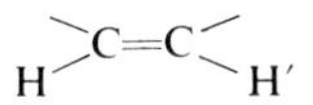 H(>C=C<)H′	(*cis*-)	6-12
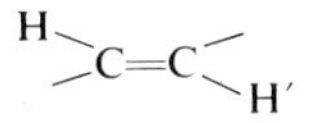 H(>C=C<)H′	(*trans*-)	13-18

1,3-coupling

—CH=C—CH′	0-3
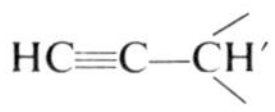 HC≡C—CH′	2-3

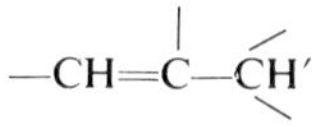

aromatic

benzene	(*ortho*-)	6-9
	(*meta*-)	1-3
	(*para*-)	0-1

PART TWO

1. Thin-layer and gas liquid chromatography

Objective

To explain the principles and practise the use of thin-layer chromatography (TLC) and of gas liquid chromatography (GLC), techniques which will be used again during this course.

Introduction

Read the section on Chromatography (pp. 15-24).

The following experiments are concerned with thin-layer chromatography (based on adsorption) and gas liquid chromatography (based on partition). The first requires very simple equipment, the second usually requires an expensive instrument.

*EXPERIMENT 1. TLC of a dye mixture**

Carry out this experiment in a fume cupboard because benzene vapour is toxic. Place benzene in a wide-neck screwtop jar (4 oz, 60 x 70 mm) to a depth of about 5 mm and fit a rectangular piece of filter paper (11 x 6 cm) around the inside of the bottle. This will become soaked with the developing solvent and ensures that the atmosphere of the jar is saturated with solvent vapour.

Your instructor will show you how to draw out a melting-point tube to make a capillary dropper. Use this dropper to apply a small spot of the dye mixture to a point 10 mm from the lower edge and 7 mm from the side of a microscope slide layered with alumina; apply a second spot 7 mm from the other side. It is important to spot the plate without damaging the thin layer and to put both spots the same distance from the lower edge. Carefully insert the slide into the developing jar, replace the cap, and screw it down lightly. It is imperative that the solvent level be *below* the sample

* Sudan IV (0.1 g), Sudan Black (0.1 g), and Rhodamine B (0.1 g) dissolved in acetone (50 ml) give a suitable dye mixture. Other mixtures are suggested in sheet 1065C-211 (*Separation of some Common Dyes*) published by Distillation Products Industries (Eastman Kodak Co.)

spots (Figure 8*c*). Allow the solvent to arise about 6 cm, remove the slide from the jar, mark the solvent front with a sharp-pointed instrument, and place the slide in a horizontal position in the fume cupboard until the solvent has evaporated.

Determine the R_f value of each component (see Figure 8*d*). Measure from the origin to the point of maximum colour intensity in each spot.

Pour the surplus solvent into the waste bottle provided (not down the sink), remove the filter paper, and allow the bottle to dry. Repeat the experiment with ether and with acetone.

Show the three slides to your instructor and report your results in the following way:

	benzene	*ether*	*acetone*
distance travelled by solvent			
distance travelled by first dye*			
distance travelled by second dye*			
etc.			

* Indicate the colour of each dye and give the R_f value in brackets for each dye and each solvent [e.g. 24 mm (0.40)].

EXPERIMENT 2. TLC of colourless compounds

The sample provided is a solution of fluorenone (200 mg), anthracene (50 mg), and diglyme (200 mg) in benzene (15 ml).

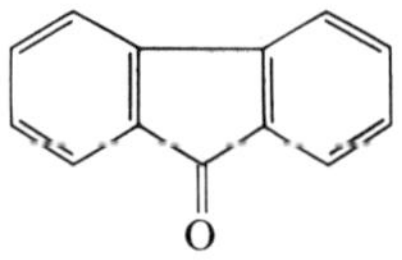

O

fluorenone
(yellow)

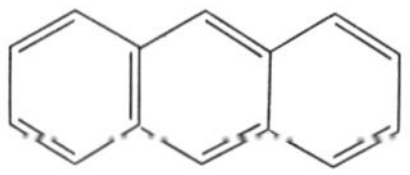

anthracene
(colourless, fluorescent)

$CH_3OCH_2 \cdot CH_2OCH_2 \cdot CH_2OCH_3$

diglyme
(colourless)

Use benzene as developing solvent for a TLC of this mixture on an alumina plate. Remember to mark the solvent front when the plate is removed from the developing jar. When the solvent has evaporated, mark the centre and outline the edge of the visible spot with a pin or sharp pencil. Next observe the plate under ultraviolet light and mark any additional spot which becomes apparent. Finally place the slide in a jar containing a few crystals of iodine and mark any new brown spots which develop after about five or ten minutes.

Record the R_f value of each spot as in *Experiment 1* and try to designate each spot from the information given about these compounds.

EXPERIMENT 3. TLC of an unknown mixture

Your instructor will give you a benzene solution containing one or more of the compounds: acetone 2,4-dinitrophenylhydrazone, azobenzene, benzophenone, and biphenyl. Samples of these individual compounds are also available in the laboratory for inspection and use. Consider how you might identify your sample by TLC (alumina plates), discuss your plans with your instructor, and then carry out the appropriate experiments.

*EXPERIMENTS 4-6. Gas Liquid Chromatography**

The following information should be recorded on each chromatogram:

Date	Stationary phase	Temperature
Sample	Mobile phase	Attenuation
	Flow rate	

The chromatogram for Experiment 5 must be run immediately after that for Experiment 4. The comparisons required in Experiment 5 can only be made if the two chromatograms are run under identical conditions.

EXPERIMENT 4. GLC of a homologous series of esters

Inject 5 μl of mixture A or B as indicated by the instructor. These are mixtures of methyl esters of monobasic (A) or dibasic (B) acids dissolved in a little ether. Have your chromatogram checked by the instructor and discuss with him what substance is represented by each peak. If the peak areas are too large or too small, inject a fresh sample of the same mixture. When you have a satisfactory chromatogram continue with *Experiment 5.* The calculations described below can be carried out later.

For a homologous series log R_t (Figure 10) is directly proportional to the number of carbon atoms and this can readily be demonstrated from your results. Measure the distance between the leading edge of the solvent peak (Figure 10) and the peak of each eluted component. This distance is directly proportional to the retention time and to

* The methyl esters of the C_8-C_{12} monobasic acids and the C_6-C_{10} dibasic acids are suitable for these experiments. They separate on ApL columns (3%) at 150-160°.

the volume of gas used to elute each component. Plot a graph of log R_t against the number of carbon atoms in each acid and satisfy yourself that these lie on a straight line. It is not usual to count the carbon atoms in the alcohol portion of the ester. From your plot calculate the retention time of some missing member of the homologous series as indicated by your instructor.

EXPERIMENT 5. Identification of methyl esters by GLC

The instructor will provide you with a sample which must contain esters of the same homologous series as was used in *Experiment 4.* This sample contains the esters of two monobasic (series A) or two dibasic acids (series B). Inject 5 μl of this mixture immediately after you have run the chromatogram in *Experiment 4* and check with the instructor that you have a satisfactory trace. This will be used for qualitative (*Experiment 5*) and quantitative analysis (*Experiment 6*). [Ideally your qualitative conclusions should be checked by re-running the unknown sample after addition of authentic samples but demand for the GLC machines may make this impossible.]

Measure R_t for the two ester peaks (Figure 10) and compare them with the results from your chromatogram in *Experiment 4.* If the mixture contains esters not present in the first mixture (A or B) see if you can get results from your log plot by interpolation or extrapolation.

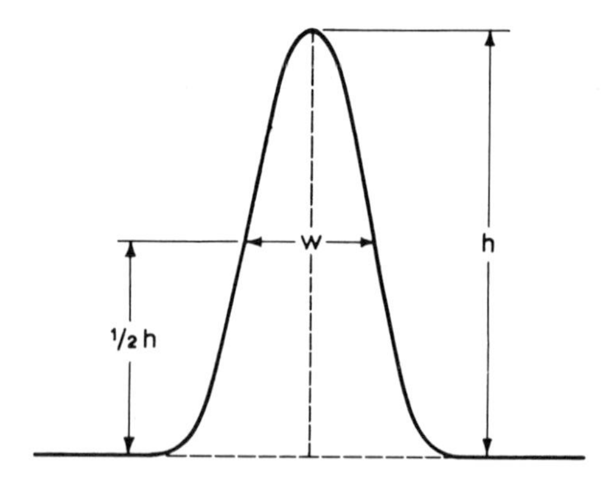

Figure 41. *GLC peak*

EXPERIMENT 6. Quantitative analysis of a mixture of two esters

Calculate the areas under the two peaks in your chromatogram (*Experiment 5*) by multiplying the height of each peak by its width at half-height (Figure 41). This gives satisfactory results with Gaussian curves. Express the results in terms of percentage (to the nearest whole number) of each component.

Experiments 5 and *6* together furnish a qualitative and quantitative analysis of a simple mixture and so illustrate two important uses of GLC.

2. Electrophilic addition to alkenes

Objectives

(1) To illustrate two important electrophilic addition reactions of alkenes (hydroxylation and bromination) and one mechanistic feature of bromination.

(2) To demonstrate the value and the limitations of hydroxylation and bromination for the detection of alkene unsaturation.

(3) To provide experience of crystallization and gas liquid chromatography.

Introduction

The alkene double bond is an electron-rich system which readily enters into addition reactions with electron-seeking (electrophilic) reagents such as potassium permanganate and bromine.

$$\text{—CH=CH—} \xrightarrow[\text{Br}_2]{\text{bromination}} \text{—CHBr·CHBr—}$$

$$\text{—CH=CH—} \xrightarrow[\text{KMnO}_4\text{, KOH}]{\text{hydroxylation}} \text{—CH(OH)·CH(OH)—}$$

Bromination of alkenes is considered to involve interaction of the double bond with a polarized bromine molecule in a stepwise process:

$$\text{—CH=CH—} \xrightarrow{\overset{\delta+\ \ \delta-}{\text{Br—Br}}} \text{—}\overset{\text{Br}^+}{\text{CH·CH}}\text{—} \begin{cases} \xrightarrow{\text{Br}^-} \text{—CHBr·CHBr—} \\ \xrightarrow{\text{X}^-} \text{—CHX·CHBr—} \end{cases}$$

This view is supported by the observation that anions other than Br^- can become involved in the second stage. In the presence of chloride ions, for example, bromination yields a mixture of dibromide and bromo-chloride as in Experiment 11.

The reaction with bromine is overall a *trans* addition. Hydroxylation by permanganate is a *cis* addition but *trans* hydroxylation can be effected with peracids such as performic acid.

Both bromination and permanganate hydroxylation are accompanied by a colour change and because it is easy to see when reaction occurs they are sometimes used as tests for alkenes. The value of these tests is limited because the reagents can react with certain other compounds which are not alkenes (see Experiment 7).

EXPERIMENT 7. Test for alkene unsaturation

The readily-observed reactions between alkenes and potassium permanganate solution or bromine solution provide a method of recognizing this functional group in organic compounds. This experiment is designed to provide practice with this test and, at the same time, to discover its limitations.

Prepare a page of your laboratory notebook as shown (p. 66). Take twelve clean test tubes (75 x 10 mm) and place ten drops (0.5 ml) of aqueous permanganate (1%) in each of six of these and ten drops (0.5 ml) of bromine in carbon tetrachloride solution (1%) in each of the remaining six. Use these to test the six compounds supplied by your instructor.

Test each substance in turn by adding one drop (liquids) or two or three small crystals (solids) of the unknown compound to the permanganate solution and to the bromine solution. Shake each tube at room temperature for one or two minutes, then record your observations. Shake vigorously if the reactants are not miscible.* When all six substances have been examined, answer question (1) and then ask your instructor for information to complete the final column in your table and answer question (2).

Questions

(1) On the basis of your observations describe each of the compounds you have examined as *an alkene*, *not an alkene*, or *uncertain*.

(2) Examine your results and compare them with what you would expect. Extend your observations by comparing results with two or three other students. Comment on the value of these two tests for alkene unsaturation. These results are

* In the permanganate test a few drops of acetone may be added if the organic compound does not dissolve.

discussed on p. 167 but this reference should not be consulted until the experiment is completed and the questions answered.

	*Permanganate solution**	*Bromine solution†*	*Name and structure of substance tested*
1			
2 etc.			

* insert: *brown ppt.* or *no reaction* as appropriate
† insert: *decolourized* or *no reaction* as appropriate

EXPERIMENT 8‡. Hydroxylation of oleic acid

$$CH_3\cdot[CH_2]_7\cdot CH{=}CH\cdot[CH_2]_7\cdot COOH \quad \text{oleic acid}$$

$$\downarrow KMnO_4$$

$$CH_3\cdot[CH_2]_7\cdot CH(OH)\cdot CH(OH)\cdot[CH_2]_7\cdot COOH \quad \text{9,10-dihydroxystearic acid}$$

Important. This reaction proceeds satisfactorily only if the quantities of reagents are carefully controlled. It is also important that the oleic acid be reasonably pure. (The mixed acids obtained by hydrolysing olive oil contain more oleic acid (*ca.* 75%) than some commercial samples of this acid).

Dissolve potassium hydroxide (2.0 g) in water (40 ml) and add this solution to oleic acid (2.0 g) contained in a conical flask (250 ml). Warm the mixture gently, with swirling, until all the acid has dissolved.

Fill a measuring cylinder with crushed ice up to the 40 ml mark, top up to this same level with cold water, and add the ice and the water to the alkaline solution of oleic acid. If the resulting solution is not at 3-5°, cool it externally with ice water. Add finely-powdered potassium permanganate (2.0 g) to the cold solution and swirl for ten minutes.

‡ Experiment 12 should be carried out before the products of Experiments 8, 9 and 10 are obtained. Read the general instructions on crystallization before starting these preparations.

Remove the excess of oxidizing agent by bubbling sulphur dioxide through the solution until it becomes colourless. Do this in a fume cupboard. Acidify with a few drops of concentrated hydrochloric acid and filter off the precipitated dihydroxy-stearic acid under suction. Use a Büchner funnel (5 cm), a filter paper to fit, and a Büchner flask (250 ml). Suck the precipitate as dry as possible, wash it with a little cold water, then again suck it as dry as possible.

The damp product is purified by crystallization. Dissolve all the product in the minimum amount of ethanol in a conical flask (25 ml) by warming the mixture on the steam bath. Allow the solution to stand at room temperature until no more crystals are deposited and then filter under reduced pressure. Crystallization may take 30-60 minutes. Place the crystals on a weighed watchglass, dry them in an oven at *ca.* 80°, and reweigh to determine the weight of product. Determine its melting point and place it in a labelled specimen tube. If the product does not have an acceptable melting point it must be crystallized again.

Write a brief account of your experiment, reporting the results as shown below, and show it, along with your specimen, to your instructor.

9,10-dihydroxystearic acid
product: wt. . . . . . . g (. . . . . .%*) m.p. . . . . . . (lit. . . . . . .)
[Give the percentage yield to the nearest whole number; 'lit.' refers to the melting point of the product quoted in the literature].

EXPERIMENT 9†. Bromination of cinnamic acid

$$C_6H_5\cdot CH{=}CH\cdot COOH \xrightarrow{Br_2} C_6H_5\cdot CHBr\cdot CHBr\cdot COOH$$

The solution of bromine in chloroform (containing 1.08 g of bromine per ml) required for this reaction should be provided and is conveniently dispensed from a burette (with a Teflon tap) kept in a fume cupboard‡. Dispense the bromine-chloroform solution (1.0 ml) into a small clean tube (75 x 10 mm) and close this with a cork.

* Your instructor will tell you how to calculate the percentage yield.
† See footnote on p. 66.
‡ Bromine is a dangerous chemical and should be handled with considerable care. If any is spilt on the skin drench the affected area with water, bathe with dilute sodium thiosulphate from the *First Aid* cabinet, and inform your instructor.

Dissolve cinnamic acid (1.0 g), contained in a conical flask (50 ml), in chloroform (5 ml) by warming gently on the steam bath. When solution is complete, swirl the flask in a beaker of water containing a little ice until small crystals just appear then immediately add the bromine solution in a single portion.

After a few minutes at room temperature the dibromide settles out. Filter this using a Hirsch funnel, filter paper of the correct size, and a filter tube. Wash the precipitate with a little cold chloroform (not more than 1 ml) and transfer it to a weighed watchglass. Reweigh when the final traces of solvent have evaporated and determine the melting point of the product.

Place the dibromide in a conical flask (25 ml) and dissolve it in the minimum quantity of aqueous ethanol (50%) by gentle warming on the steam bath. Allow the clear solution to cool to room temperature then filter the crystallized product and determine its weight and melting point as before.

Write a brief account of your experiment, reporting the results as shown below, and show it, along with the product, to your instructor.

2,3-dibromo-3-phenylpropanoic acid
crude product: wt. ...... g (...... %)* m.p. ......
crystallized product: wt. ...... g (...... %)* m.p. ...... (lit. ......)

[Give the percentage yield to the nearest whole number; 'lit.' refers to the melting point of the product quoted in the literature].

EXPERIMENT 10.† *Bromination of styrene*

$$C_6H_5 \cdot CH{=}CH_2 \xrightarrow{Br_2} C_6H_5 \cdot CHBr \cdot CH_2Br$$

Bromination of styrene (phenylethylene) is effected by a solution of bromine (34 g) in carbon tetrachloride (50 ml). This is conveniently dispensed from a burette (with a Teflon tap) kept in a fume cupboard.‡

Measure the bromine solution (5 ml) into a conical flask (25 ml) and carefully add styrene (2.2 g, 2.0 ml) also from a burette. Mix and allow to stand at room temperature for a few minutes. Cool the mixture in ice, and when crystallization is

* See footnote on p. 67.
† See footnote on p. 66.
‡ See footnote on p. 67.

complete, quickly filter the suspension using a Hirsch funnel, a filter paper of the correct size, and a filter tube. Keep a little of the solid for m.p. determination and crystallize the remainder from 80% ethanol. Place the crude product in a small conical flask, just cover it with 80% ethanol, and heat on a steam bath until it dissolves. If necessary, add a little more solvent. Allow the hot solution to cool to room temperature and, if possible, leave it for one hour before filtering as already described. Wash the crystals with a small amount of cold solvent then transfer them to a weighed watchglass. Dry the sample by placing it in a desiccator overnight. Determine the weight and melting point of the styrene dibromide (1,2-dibromo-1-phenylethane)[1].

Write a brief account of your experiment, reporting the results as shown below, and show it, along with your product, to your instructor.

1,2-dibromo-1-phenylethane
product: wt. ...... g (...... %)* m.p. before purification ......
m.p. after crystallization ..... (lit. ......)

[Give the percentage yield to the nearest whole number; 'lit.' refers to the melting point of the product quoted in the literature]

Note [1] Some people find this compound to be a skin irritant; it should therefore be handled with great care.

EXPERIMENT 11. Bromination of oct-1-ene in the presence of lithium chloride

Prepare a solution of bromine (5.4 g) in acetic acid (100 ml) and to *one half* of this add anhydrous lithium chloride (0.72 g). This solution should be kept in the dark. It may lose strength if kept too long.

(*a*) Add oct-1-ene (0.5 ml) to the bromine solution (10 ml) contained in a stoppered flask (20 ml) and shake vigorously for five minutes. Transfer the mixture to a separating funnel[1] containing 10% sodium thiosulphate solution (50 ml). Rinse out the reaction flask with ether (10 ml) and transfer this also to the separating funnel. Shake well, allow the two layers to separate, and run off the lower aqueous layer. Wash the ethereal extract with water (10 ml) and dry the extract over anhydrous magnesium sulphate. Filter the dried solution through a Hirsch funnel using gentle suction. This solution contains 1,2-

* See footnote on p. 67.

dibromo-octane and should be kept in a well-stoppered flask until it is examined by GLC[2].

(*b*) Repeat the above experiment using the bromine solution which also contains lithium chloride.

(*c*) Carry out a GLC examination[2] of the product from experiment (*a*) and from experiment (*b*). Identify the peaks in each chromatogram by comparison with chromatograms of authentic substances. (Read p. 23 about the identification of compounds by GLC).

Notes [1] If you do not know how to use a separating funnel, consult your instructor.

[2] Recommended GLC conditions: 3% ApL at 150°.

Questions (1) Write equations for experiments (*a*) and (*b*).

(2) What other product might be formed in experiment (*a*)? Is there any evidence for this in your chromatogram?

(3) Would you expect 1-bromo-2-chloro-octane or 2-bromo-1-chloro-octane or both these isomers to be produced in experiment (*b*)? Give a reason for your answer. How would you determine this experimentally?

EXPERIMENT 12. Melting point and mixed melting point

Read the section on Melting Point on p. 8.

Your instructor will show you how to seal and fill a melting-point tube and how to take a melting point and mixed melting point.

You are provided with three samples (A, B, and C) all of which have similar melting points. Determine the melting range of (i) substance A, (ii) a mixture of substances A and B, and (iii) a mixture of substances A and C.

Record these in your laboratory book and indicate whether A is identical with B, with C, or with neither of these substances. Check your results and your conclusions with your instructor.

3. Nucleophilic addition to carbonyl compounds

Objectives

(1) To discover to what extent aldehydes and ketones can be detected by:
  (*a*) reaction with 2,4-dinitrophenylhydrazine,
  (*b*) their infrared spectra.

(2) To identify an unknown aldehyde or ketone through the melting point of its 2,4-dinitrophenylhydrazone and/or semicarbazone.

(3) To recognize these reactions as typical addition-elimination reactions of carbonyl compounds.

(4) To provide further experience of crystallization.

Introduction

Read this section carefully before undertaking any of the following experiments. Insert appropriate words or structures where a gap is indicated by —— and have your instructor check these before you start the experimental section.

The carbonyl group is $>C{=}O$ and any compound containing this group can be described as a carbonyl compound. These fall into two main classes: aldehydes and ketones on the one hand and carboxylic acids and their derivatives on the other.

Below are formulated an acid anhydride, acyl chloride, aldehyde, amide, anion of a carboxylic acid, carboxylic acid, ester and ketone. Assign these names to the appropriate structures.

$R{-}C(=O){-}H$　　$R{-}C(=O){-}O^-$　　$R{-}C(=O){-}OR'$

$R{-}C(=O){-}R$　　$R{-}C(=O){-}Cl$　　$R{-}C(=O){-}NH_2$

$R{-}C(=O){-}OH$　　$R{-}C(=O){-}O{-}C(=O){-}R$

Reactions of aldehydes and ketones with nucleophilic reagents. The characteristic reactions of aldehydes and ketones are *addition* reactions occurring at the unsaturated carbonyl (C=O) group.

$$\text{e.g.}\quad \underset{\text{aldehyde}}{R{-}C{\lesssim}^{O}_{H}} \xrightarrow[H_2O]{KCN} \underset{\text{cyanohydrin}}{R{-}\overset{OH}{\underset{H}{C}}{-}CN}$$

$$R{-}C{\lesssim}^{O}_{R} \xrightarrow[H_2O]{KCN} \text{——}$$

Sometimes the *addition* reaction is followed by an *elimination* and the whole reaction is an *addition-elimination* process. The conversion of aldehydes and ketones to oximes by reaction with hydroxylamine is an example of an addition-elimination reaction. So also is the formation of a hydrazone.

$$\underset{\text{hydroxylamine}}{R\cdot CH{=}O + NH_2OH} \xrightarrow{\text{addition}} \left[R\cdot\overset{OH}{\overset{|}{C}}H{-}NHOH\right] \xrightarrow[-H_2O]{\text{elimination}} \underset{\text{an oxime}}{R\cdot CH{=}NOH}$$

$$\underset{\text{hydrazine}}{R_2C{=}O + NH_2NH_2} \rightleftharpoons \left[R_2\overset{OH}{\overset{|}{C}}{-}NHNH_2\right] \xrightleftharpoons[-H_2O]{} \underset{\text{a hydrazone}}{R_2C{=}NNH_2}$$

2,4-Dinitrophenylhydrazine (Brady's reagent) is an important reagent related to hydrazine. This compound, with the structure shown, is conveniently represented as $ArNHNH_2$ (Ar is the symbol for an aryl or aromatic group). Most aldehydes and ketones react very readily with this reagent to give 2,4-dinitrophenylhydrazones and this reaction is used in two ways:

$NHNH_2$

O_2N NO_2

2,4-dinotrophenylhydrazine

(*a*) A positive reaction, shown by the formation of a yellow, orange, or red precipitate, indicates the presence of an aldehyde or ketone.

(*b*) The dinitrophenylhydrazone can be isolated, purified by crystallization, and its melting point used to identify the aldehyde or ketone.

$$R_2C{=}O + NH_2NHAr \longrightarrow \text{———} \xrightarrow[-H_2O]{} R_2C{=}NNHAr$$

Aldehydes and ketones can also be characterized as **oximes** or **semicarbazones.**

$$R_2C{=}O + NH_2NHCONH_2 \longrightarrow R_2C{=}NNHCONH_2$$

a semicarbazone

Carboxylic acids and their derivatives do not undergo reactions of this type. They do not, for example, form phenylhydrazones even though they contain a carbonyl group.

Infrared spectra of carbonyl compounds. Practically all carbonyl compounds (aldehydes, ketones, and the carboxylic acids and their derivatives) show strong absorption in the 1650-1850 cm^{-1} region of the infrared. The exact position of this absorption band may indicate the specific type of carbonyl group. At this stage we are concerned only to recognize a carbonyl compound from its infrared spectrum without further designation. (For further details and discussion see p. 30). Take care not to confuse carbonyl absorption with the absorption around 1600 cm^{-1} shown by some aromatic compounds.

EXPERIMENT 13. Detection of the carbonyl group by infrared spectroscopy and by reaction with 2,4-dinitrophenylhydrazine.

(*a*) On a fresh sheet of infrared chart paper clearly mark the region of carbonyl absorption (1650-1850 cm^{-1}) by shading it in. Fasten the paper into your laboratory notebook. Further insertions into this chart will be made in Experiment 25.

(*b*) Draw up, in your notebook, a table like that on the following page. Examine the infrared spectra (pp. 38-47) of the substances listed and insert in your table the wavenumber of any carbonyl absorption band(s). Where necessary write in the word *none*.

(*c*) In each of eight clean dry test tubes (75 x 10 mm) put 2,4-dinitrophenylhydrazine solution* (1 ml). Test each substance by adding *one drop* of the liquid (you may get erroneous results if you add more than this) or two or three crystals of solid previously dissolved in the minimum volume of ethanol. If there is no precipitate on shaking, boil for one minute and cool. Report your results as *no precipitate* or as a(n) *yellow/orange/red precipitate.*

Substances to be examined		*Infrared maxima* (cm^{-1})	*2,4-dinitrophenylhydrazone*
Structure	*Name*		
$C_6H_5 \cdot NH \cdot CO \cdot CH_3$	acetanilide		
$C_6H_5 \cdot OCH_3$	anisole		
$C_6H_5 \cdot CHO$	benzaldehyde		
$C_6H_{10}=O$	cyclohexanone		
$CH_3 \cdot CO \cdot OC_2H_5$	ethyl acetate		
$C_6H_5 \cdot CO \cdot CH_2 \cdot CH_2CO \cdot OC_2H_5$	ethyl 4-oxo-4-phenylbutanoate		
$CH_3 \cdot [CH_2]_5OH$	hexanol		
$CH_3 \cdot [CH_2]_5 \cdot CO \cdot CH_3$	octan-2-one		

* 2,4-Dinitrophenylhydrazine reagent is made by dissolving 2,4-dinitrophenylhydrazine (40 g) in concentrated sulphuric acid (80 ml). This solution is cooled, added to methanol (900 ml), and finally diluted with water (100 ml). The reagent is decanted from any crystalline solid which deposits on standing.

Having completed the table, classify your substances into three categories: (*a*) those showing carbonyl absorption in the infrared and giving a precipitate with dinitrophenylhydrazine, (*b*) those showing carbonyl absorption which do not give a precipitate, and (*c*) those not showing carbonyl absorption and not giving any precipitate. Which of these are aldehydes and ketones, which are carbonyl compounds other than aldehydes and ketones, and which are not carbonyl compounds?

EXPERIMENT 14. Identification of an aldehyde or ketone from the melting point of its 2,4-dinitrophenylhydrazone and/or semicarbazone.
Prepare the 2,4-dinitrophenylhydrazone and/or semicarbazone of an unknown aldehyde or ketone. Crystallize the product, determine its melting point, then recrystallize it and measure the melting point again. Place the derivative in a labelled specimen tube and show it to your instructor. Write out the equation for the reaction you carry out.

Consult tables of melting points and try to discover the identity of the aldehyde or ketone. If you cannot decide between several possibilities, list them all and give the appropriate melting points. If possible, suggest how you would complete the identification. If your unknown is a solid, determine its melting point also. For this experiment only one derivative need be prepared but in identifying an unknown compound at least two points of identity are generally required.

EXPERIMENT 14a. Preparation of a 2,4-dinitrophenylhydrazone.
Place the unknown compound (400 mg) in a Quickfit test tube and dissolve it in the minimum quantity of ethanol if it is solid. Add 2,4-dinitrophenylhydrazine solution (20 ml)* and shake the tube to mix the contents thoroughly. If a precipitate appears on mixing, keep the mixture at room temperature for 20 minutes; if a precipitate does not appear immediately, attach a reflux condenser, add two or three anti-bumping granules, reflux on the steam bath for 5 minutes, and cool. Filter off the precipitate using a Hirsch funnel and filter tube and wash the precipitate with a little cold ethanol.

The derivative must now be purified by crystallization from ethanol or acetic acid. (Do not use acetic acid unless the derivative is insufficiently soluble in ethanol.) Transfer the product to a conical flask (100 ml), add a little ethanol, and reflux the mixture. If the solid does not dissolve after a few minutes, add more solvent in small

* See footnote on p. 74.

portions and continue to reflux until solution is complete. (If the product does not dissolve in 50 ml of solvent, consult your instructor.) Allow the clear solution to cool and leave it at room temperature for 20-30 minutes before filtering with a Hirsch funnel. Dry the product thoroughly and determine its melting point. Repeat the crystallization and melting-point determination.

Examine a list of melting points and report any which resemble that of the unknown:

melting point of derivative
after first crystallization
after second crystallization
possible aldehydes/ketones (list names and give m.p.)
Write an equation for the reaction you have carried out.

EXPERIMENT 14b. Preparation of a semicarbazone.
Prepare a solution of semicarbazide hydrochloride (0.5 g) and hydrated sodium acetate (0.75 g) in water (2 ml) in a conical flask (50 ml) and add the carbonyl compound (0.5 g: if your unknown is a solid, dissolve it in the minimum amount of ethanol). If the carbonyl compound does not dissolve, add a few drops of ethanol but do not add a large volume of alcohol or sodium chloride will be precipitated. If a derivative does not separate in 5 minutes, add two or three anti-bumping granules, attach a reflux condenser, reflux on the steam bath for 5-10 minutes (not more), and cool. If much ethanol has been used it may be necessary to concentrate the solution before the derivative will crystallize. Finally the product is filtered, using a Hirsch funnel and filter tube, and the precipitate washed with a little cold water.

Recrystallize the semicarbazone from ethanol as described in *Experiment 14a.* Report your results as shown in the previous *Experiment* (*14a*) and write an equation for the reaction you have carried out.

4. Nucleophilic substitution of saturated compounds

Objectives

(1) To illustrate the varying reactivities of alkyl and aryl halides.

(2) To prepare an alkyl halide as an illustration of a nucleophilic substitution at a saturated carbon atom.

(3) To provide experience in the use of a separating funnel and of distillation.

(4) To provide further experience of gas liquid chromatography and to illustrate the usefulness of this technique.

(5) To provide experience of column adsorption chromatography as a means of purification.

Introduction

Many reactions which at first sight appear to be diverse are in fact substitution reactions occurring at a saturated centre by a common mechanism. The attacking reagent is usually nucleophilic and the reaction can be represented by the general expression:

$$N^- + C—L \longrightarrow N—C + L^-$$

where N^- is the nucleophilic reagent, C the saturated carbon atom, and L^- the displaced or leaving group. N and L may or may not carry a charge.

EXPERIMENT 15. Solvolysis of alkyl halides

The word solvolysis is used to describe a reaction in which the attacking reagent also serves as solvent. The reaction between an alkyl halide (RX) and a protonic solvent (HS) can be written:

$$HS + RX \longrightarrow RS + H^+ + X^-$$

This experiment is designed to show how the reaction is affected by changes of R and X. These effects are completely demonstrated only by a careful determination of the kinetics of the solvolysis of a range of halides but a simpler qualitative approach is used here. The reactions under study are unimolecular processes (S_N1) in which the rate-determining step is the ionization of the alkyl halide. The carbonium ion so produced reacts rapidly with a nucleophile (R′OH or H_2O) or loses a proton to give an olefin:

$$R{-}X \longrightarrow X^- \text{ and } R^+ \begin{cases} \xrightarrow{R'OH} ROR' + H^+ \\ \xrightarrow{H_2O} ROH + H^+ \\ \longrightarrow \text{alkene} + H^+ \end{cases}$$

Although the reaction described below is catalysed by silver ions added to the reaction mixture for a different purpose, the formation of the carbonium ion remains the rate-determining step. The reaction is followed by noting the rate at which silver halide is precipitated.

$$R{-}X \xrightarrow[\text{slow}]{Ag^+} AgX + R^+ \xrightarrow[\text{fast}]{} \text{products}$$

The silver nitrate solution required contains silver nitrate (0.5 g) dissolved in a mixture of water (50 ml) and ethanol (50 ml). Place this solution (20 ml) in each of three conical flasks (50 ml). Pipette the halogen compounds (0.2 ml) indicated below into separate small sample tubes and transfer them simultaneously to the flasks containing silver nitrate solution. Cork the flasks, shake them, allow to stand for 30 minutes, and observe whether any precipitate of silver halide appears. At the end of this time warm on the steam bath any flask in which little or no precipitate has appeared and continue to watch for a precipitate for a further 30 minutes. Repeat the experiment with each set of halides. Report and comment on your results and discuss them with your instructor. Be sure to write out the structures of all the halides.

(*a*) 1-chlorobutane (primary), 2-chlorobutane (secondary), and 2-chloro-2-methylpropane (tertiary)
(*b*) 2-chlorobutane, 2-bromobutane, and 2-iodobutane
(*c*) 1-bromobutane, allyl bromide, and bromobenzene.

EXPERIMENTS 16-18. The preparation of halogenoalkanes from primary, secondary, and tertiary alcohols

Write the following equation in your notebook replacing R by the appropriate alkyl group. If you do not understand it, discuss it with your instructor. Prepare, isolate, and distil one of these halides and check its purity by GLC. Pay careful attention to

$$X^- + \underset{\text{protonated alcohol}}{R{-}\overset{+}{O}H_2} \longrightarrow X{-}R + H_2O$$

the notes appended to each set of experimental details.

Primary alcohols are usually reacted in the presence of sulphuric acid, probably because the sulphate so formed contains a better leaving group. Sulphuric acid is not necessary with secondary and tertiary alcohols and may in fact enhance the competitive elimination reaction.

EXPERIMENT 16. Preparation of 1-bromopentane

Read the introduction to Experiments 16-18 and the general instructions on distillation (p. 9) before starting this experiment. Two methods for preparing this alkyl halide are given using (i) hydrobromic acid and sulphuric acid and (ii) sodium bromide and sulphuric acid. The recovery procedure is the same in each case.

(i) *Preparation using hydrogen bromide.* Place hydrobromic acid (48%, 62 g, 42 ml) in a round-bottomed flask (250 ml) and swirl it in an ice-water bath whilst slowly adding concentrated sulphuric acid (16 g, 9 ml). Then carefully add pentan-1-ol (25 g, 31 ml) and a few anti-bumping granules, fit a condenser, and reflux the mixture over a Bunsen burner in a fume cupboard for one hour. The product is isolated as indicated in (iii).

(ii) *Preparation using sodium bromide.* Place sodium bromide (36 g), water (33 ml), and pentan-1-ol (25 g, 31 ml) in a round-bottomed flask (250 ml) and cool the mixture to below 5° in an ice-water bath. Add concentrated sulphuric acid (44 g, 24 ml) slowly with continuous swirling and cooling. Add a few anti-bumping granules, insert a water condenser, and boil the mixture over a Bunsen burner in a fume cupboard for one hour. Isolate the product as indicated in the following section.

(iii) *Isolation of bromopentane.* Add water (50 ml) to the reaction mixture and after making the apparatus suitable for distillation[1], distil until no more oily distillate appears[2]. Pour the distillate into a separating funnel (250 ml) containing water (50 ml) and rinse the flask with ether (2 x 30 ml)[3] adding each portion of ether to the separating funnel[3]. After shaking the funnel and separating the two layers the ether solution is washed with water (20 ml), dilute sodium hydroxide (2 *M*, 20 ml), and again with water (20 ml). Finally run the ether layer into a dry conical flask containing anhydrous sodium sulphate or magnesium sulphate (*ca.* 5 g). Close the flask and leave it for 20 to 30 minutes.

While the solution is drying, set up a distillation apparatus consisting of a flask (250 ml), distillation head, thermometer, condenser, receiver adapter and receiver. This last should be a 250 ml flask for the solvent and *weighed* 50 ml flasks for any intermediate fraction and for the product. All the apparatus must be dry. Have the apparatus checked by your instructor before use and make sure that it is suitable for ether distillation.

Filter the ether solution of 1-bromopentane from the drying agent by pouring it through a filter funnel which has a small plug of cotton wool pressed into the upper part of the stem. Wash out the flask and funnel with a little more ether (5-10 ml), add two or three anti-bumping granules, and distil off the solvent[4] on a steam bath. Then transfer the residual liquid to a smaller flask (100 ml) and distil the bromopentane using an oil bath. Place a thermometer in the oil bath and raise the bath temperature until the bromopentane (b.p. 130°) begins to distil[5]. Keeping the bath temperature steady, collect the distillate in a weighed receiver, and note the distillation temperature range over which it is collected.

(iv) *GLC examination.* Examine the distilled product by GLC[6]. This should show bromopentane as the major product, accompanied by some byproducts, and by unreacted pentanol. The bromide is easily purified by adsorption chromatography[7].

(v) *Purification by adsorption chromatography.* (see p. 15). Pack a chromatography column (*ca.* 30 x 2 cm) with alumina (*ca.* 20 cm high) in petroleum (b.p. 40-60°). Run off surplus solvent until the top of the alumina is just covered, then pipette some of your bromopentane (*ca.* 1 g) on to the column and run off more petroleum until the liquid level again just covers the alumina. Top up the column with petroleum and elute with this solvent until the bromopentane has come off the column (*ca.* 100 ml of solvent should be sufficient). The more polar pentanol is more strongly adsorbed by the alumina and is eluted only with a more polar solvent such as diethyl ether.

Carefully distil the petroleum from the eluate on a steam bath and check the purity of the product by GLC. (Both GLC and TLC are convenient means of monitoring column chromatography experiments).

(vi) *Results.* Write a brief account of your experiment and report the results as follows:

1-bromopentane
distillate: wt. ...... g (..... %), boiling range ......
Chromatographic purification: purity before chromatography ...... %
purity after chromatography ...... %
Include the *two* GLC traces in your notebook.

Notes

[1] If you do not know how to do this consult your instructor.
[2] This is a simple form of steam distillation in which steam is generated by boiling an aqueous mixture. Do not allow the volume of liquid in the distilling flask to become too small. If necessary more water can be added and distillation continued but do this carefully after *cooling* as the residual liquid will be strongly acidic. Exercise the same caution when washing out the flask at the end of the experiment.
[3] If you have not used a separating funnel before, ask your instructor to show you how to handle it.
[4] Put unwanted ether distillates into the vessel provided. *DO NOT throw them down the sink.*
[5] When distilling from an oil bath, the bath temperature usually has to be 30-50° above the boiling point of the distillate.
[6] Use a column of celite coated with PEGA (20%) at a temperature of 95°.
[7] In most instructions for preparing alkyl halides from alcohols the product is shaken with concentrated sulphuric acid to remove unreacted alcohol. This operation is potentially dangerous. The chromatographic procedure illustrates a valuable general technique.

EXPERIMENT 17. Preparation of 2-bromopentane

Read the introduction to Experiments 16-18 and the general instructions on distillation (p. 9) before starting this experiment.

Reflux a mixture of pentan-2-ol (25 g, 31 ml) and hydrobromic acid (48%, 62 g, 42 ml) in a round-bottomed flask (250 ml) containing a few anti-bumping granules, for 30 minutes over a Bunsen burner in a fume cupboard. Isolate the bromopentane (b.p. 117°) as described in section (iii) of *Experiment 16* and examine your distilled product by GLC (see *Experiment 16*, section iv). If it contains unchanged alcohol, purify about 1 g by adsorption chromatography (*Experiment 16*, section v) and check its purity by GLC.

Report your results as shown in *Experiment 16* (section vi).

EXPERIMENT 18. Preparation of 2-chloro-2-methylpropane (t-butyl chloride)

Read the introduction to *Experiments 16-18* and the general instructions on distillation (p. 9) before starting this experiment.

Place 2-methylpropan-2-ol (t-butanol, 25 g, 32 ml) in a separating funnel (250 ml), add concentrated hydrochloric acid (85 ml), and shake these together for a few minutes. This operation should be done with great care. Wear a face shield and rubber gloves and do it away from other unprotected persons.

Leave the mixture for about 20 minutes when the two layers should separate [1]. Run off the lower aqueous layer and wash the upper layer with water (25 ml), with sodium bicarbonate solution (0.5 *M*, 25 ml; see Note [2] which is very important), and again with water (25 ml). Run the product into a dry conical flask containing anhydrous sodium sulphate or magnesium sulphate (*ca*. 5 g.).

Set up a distillation apparatus consisting of a flask (100 ml), distillation head, thermometer, condenser, receiver adapter, and weighed receiver flask (two 50 ml flasks may be needed). All the apparatus must be dry. Have your apparatus checked by your instructor before use. After *ca.* 20 minutes filter the t-butyl chloride from the drying agent, by pouring it through a filter funnel which has a small plug of cotton wool pressed into the upper part of the stem into the distillation flask containing a few anti-bumping granules. Distil the t-butyl chloride, heating the flask with an oil bath equipped with a thermometer[3]. Examine the distillate by GLC[4] to see if it contains t-butanol. Report your results in the following way:

2-chloro-2-methylpropane
distillate: wt. ...... g (...... %), boiling range ......

Notes

[1] If the two layers are slow to separate, the alkyl chloride can be salted out by adding calcium chloride until the aqueous layer is saturated.

[2] Ask your instructor to show you how to vent the carbon dioxide during this operation: failure to do this could cause your separating funnel to explode.

[3] When distilling from an oil bath the bath temperature usually has to be 30-50° above the distillation temperature. The boiling point of t-butyl chloride is 51°.

[4] Use a celite column coated with PEGA (20%) at a temperature of 50°.

*EXPERIMENT 19. Competitive reaction of butan-2-ol with hydrochloric and hydrobromic acids**

$$CH_3\cdot CH_2\cdot \underset{\underset{^{+}OH_2}{|}}{CH}\cdot CH_3 \begin{cases} \xrightarrow{Cl^-} CH_3\cdot CH_2\cdot CHCl\cdot CH_3 \\ \xrightarrow{Br^-} CH_3\cdot CH_2\cdot CHBr\cdot CH_3 \end{cases}$$

Copy this equation into your notebook and name the two products. If you do not understand the equation discuss it with your instructor. In this experiment a secondary alcohol is treated with a mixture of halogen acids. The reaction product is isolated, analysed by GLC, and an attempt made to separate the two components by distillation.

Preparation. In a round-bottomed flask (250 ml) place butan-2-ol (37 ml, 0.04 mole), concentrated hydrochloric acid (42 ml, 0.4 mole), and concentrated (47%) hydro-

* This experiment is based on a similar one from Helmkamp and Johnson's *Selected Experiments in Organic Chemistry* and we are indebted to them and their publisher (W. H. Freeman) for permission to reproduce our modification of their experiment.

bromic acid (70 ml, 0.4 mole) and then carefully add concentrated sulphuric acid (20 ml) with swirling. Add two or three anti-bumping granules, fit a reflux condenser, and reflux over a Bunsen burner in the fume cupboard for 45 minutes.

Cool the reaction mixture and pour it into a separating funnel (500 ml)[1] containing crushed ice (200 g); rinse out the reaction flask with a little cold water (10-20 ml) and add the washings to the separating funnel. Swirl the funnel and allow the two layers to settle and all or most of the ice to melt. Transfer the organic layer to a second separating funnel (250 ml) and wash with water (50 ml), sodium hydroxide solution (2 *M*, 2 x 50 ml), and water (50 ml). Now put the organic layer into a dry conical flask (100 ml) containing anhydrous sodium sulphate or magnesium sulphate (5-10 g) and let it stand for 20-30 minutes.

Whilst this is drying, set up a distillation unit with dry apparatus. This consists of a distillation flask (100 ml), stillhead, condenser, receiver adapter, and two weighed receiver flasks (25 ml). Use an oil bath to heat the flask and have a thermometer in the oil bath[2]. Have your apparatus checked by the instructor before you use it and consult him about any doubtful points.

Filter the dried product by passing it through a filter funnel containing a little cotton wool pressed into the upper part of the stem. Drain the flask and funnel as thoroughly as possible. Before distilling remove about 1 ml of the product for GLC examination and add two or three anti-bumping granules to the remainder.

The chloride boils at 68° and the bromide at 91° but these are not well separated in a simple distillation apparatus. If there is no apparent break in the distillation rate as the temperature rises, collect two fractions of about equal volume, note the boiling range of each, and weigh them.

GLC examination. Examine samples of the total product, the distilled fractions, and the distillation residue *under the same chromatographic conditions*[3]. Indicate what components are present in each sample and from the areas of the chloride and bromide peaks compute the proportion of these in each fraction. Tabulate your results, as shown below, along with a brief account of the experiment and attach the chromatograms to your notebook.

From the results on the total product calculate the molar yield of chloride and bromide. If these are not the same try to explain the discrepancy.

	Boiling range	*Wt (g)*	*Chloride (%)*	*Bromide (%)*
total product	–	–		
fraction one				
fraction two				
residue	–	–		

Comment on the efficiency of the distillation and suggest how it might be improved.

Notes [1] If you have not used a separating funnel before, ask your instructor to show you how to handle it.

[2] When distilling from an oil bath the bath temperature usually has to be 30-50° above the distillation temperature.

[3] Recommended GLC conditions: 20% PEGA on celite; temperature 65°.

5. Substitution reactions of acyl compounds

Objectives

(1) To carry out some reactions of carboxylic acids and esters as examples of substitution reactions at an unsaturated carbon atom.

(2) To illustrate the use of acylation procedures as methods of characterizing amines, alcohols, and phenols, and also carboxylic acids.

Introduction

Many reactions of carboxylic acids and their derivatives (acid chlorides, esters, anhydrides, amides) can be considered as substitution reactions occurring at an unsaturated carbon atom. For example, the reaction between ammonia and an ester to form an amide can be written:

$$\underset{\text{ester}}{R \cdot COOEt + NH_3} \longrightarrow \underset{\text{amide}}{R \cdot CONH_2 + EtOH}$$

To emphasize the idea of substitution we can write the following (simplified) mechanism:

$$H_3\ddot{N} \; \underset{R}{\overset{O}{\gg}} C{-}OEt \longrightarrow H_3\overset{+}{N}{-}C\underset{R}{\overset{O}{\lessgtr}} + \bar{O}Et \xrightarrow[\text{transfer}]{\text{proton}} H_2N{-}C\underset{R}{\overset{O}{\lessgtr}} + EtOH$$

This is the mechanistic theme behind most of the reactions in this exercise.

Carboxylic acids and their derivatives can be represented by the general formula

$R \cdot C\underset{Z}{\overset{O}{\lessgtr}}$ Where Z is OH (carboxylic acid), NH_2 (———), OR′ (———), Cl (———), or OCO·R (———).

The R · CO group is an *acyl* group and the introduction of this group into a molecule is *acylation.* The most important acylating agents are acyl halides (R · COCl)

and acid anhydrides (R.COOCO.R). The compounds most readily acylated contain O—H groups (alcohols and phenols) or N—H groups (amines) and acylation reactions are often used to characterize such compounds. The reaction can be used in the opposite sense to characterize unknown acids. For example:

acylation of an alcohol or amine:

$$R'\cdot COCl \begin{cases} \xrightarrow{ROH} R'\cdot COOR \text{ (ester)} \\ \xrightarrow{RNH_2} R'\cdot CONHR \text{ (substituted amide)} \end{cases}$$

characterization of a carboxylic acid via its acid chloride:

$$R'\cdot COOH \longrightarrow R'\cdot COCl \begin{cases} \xrightarrow{NH_3} R'\cdot CONH_2 \text{ (amide)} \\ \xrightarrow{RNH_2} R'\cdot CONHR \text{ (substituted amide)} \end{cases}$$

These reactions can be formulated in the mechanistic terms which have already been discussed. Complete the following sequences:

$$R\ddot{N}H_2 \quad R'(O=)C{-}Cl \longrightarrow \underline{\qquad} + \underline{\qquad} \xrightarrow[\text{transfer}]{\text{proton}} RNH{-}C(=O)R' + HCl$$

$$R\ddot{O}H \quad R'(O=)C{-}Cl \longrightarrow \underline{\qquad} + \underline{\qquad} \xrightarrow[\text{transfer}]{\text{proton}} \underline{\qquad} + HCl$$

EXPERIMENT 20. Preparation of methyl benzoate

Methyl benzoate is prepared by the acid-catalysed reaction between benzoic acid and methanol. In simplified form we can represent the reaction:

$$\begin{array}{ccc} Ph\cdot COOH & & Ph\cdot COOMe \\ \updownarrow H^+ & & \updownarrow H^+ \\ (Me)(H)\ddot{O}: \; + \; Ph{-}C(=O){-}\overset{+}{O}H_2 & \rightleftharpoons & (Me)(H)\overset{+}{O}{-}C(=O)Ph + H_2O \end{array}$$

benzoic acid (protonated) — methyl benzoate (protonated)

Place benzoic acid (12.2 g) in a round-bottomed flask (250 ml) and add methanol (40 ml). Swirl the contents of the flask whilst carefully adding concentrated sulphuric acid (5 ml), and wash any sulphuric acid off the inside neck of the flask with a little more methanol (10 ml). Add two or three anti-bumping granules to the mixture and fit a reflux condenser.

Reflux this mixture on the steam bath for one hour, and after cooling pour the reaction mixture into a separating funnel (500 ml) containing water (150 ml). Rinse the reaction flask with carbon tetrachloride (2 x 40 ml) and add this to the funnel. Use this organic solvent to extract the ester, remove the aqueous layer, and wash the carbon tetrachloride solution with water (30 ml). Dry the ester solution with anhydrous sodium sulphate and filter the solution. Distil off the solvent on a steam bath, then transfer the residual ester to a smaller flask (50 ml) and distil using a Bunsen burner (b.p. 198°). Run the water out of the condenser when distilling the ester. Collect the ester in a weighed flask and note the distillation temperature range.

Write a brief account of your experiment and report the result as follows:

methyl benzoate ...... g, (...... %), b.p. ......

Question. The ester solution is washed with alkali to remove unreacted benzoic acid. Given a two-phase mixture of water and carbon tetrachloride, in which phase would the following concentrate: (*a*) methyl benzoate, (*b*) methanol, (*c*) sulphuric acid, (*d*) benzoic acid and (*e*) sodium benzoate?

*EXPERIMENT 21. Hydrolysis of an unknown ester**

Hydrolyse the ester provided, isolate the resulting acid which will be water-insoluble, and try to identify it from its melting point and the GLC behaviour of the original ester.

$$\bar{HO} + R{-}C({=}O){-}OR' \longrightarrow HO{-}C({=}O){-}R + \bar{O}R' \xrightarrow[\text{exchange}]{\text{proton}} \bar{O}{-}C({=}O){-}R + R'OH$$

Place the ester (5 g) in a round-bottomed flask (100 ml) fitted with a reflux condenser and Teflon sleeve[1]. Add sodium hydroxide solution (2 *M*, 50 ml) and a few anti-bumping granules, and clean the neck of the flask with a wet filter paper[2].

* These directions are suitable only for esters of water-soluble alcohols and solid water-insoluble acids.

Attach a reflux condenser and reflux for 30-60 minutes. When hydrolysis is complete, the organic products will be completely soluble in the aqueous alkali. Cool the reaction mixture, carefully acidify with concentrated hydrochloric acid, and filter off the white precipitate using a Büchner funnel and filter flask. Wash the acid with a little cold water, press it as dry as possible, then dry it on a weighed watchglass in the oven†. Report the weight and melting point of your product. Recrystallize a sample of the acid and re-determine its melting point.

Consult a list of melting points to try to identify your acid. Your instructor will inform you what the alcohol component of your ester is. Check your conclusion by GLC. Examine, under the same experimental conditions[3]: (*a*) your unknown ester, (*b*) an authentic sample of the ester you think you have (supplied by the instructor), and (*c*) a mixture of the two made from two or three drops of each. Attach the chromatographs to your notebook and record your conclusions.

Notes
[1] Consult your instructor about this.
[2] It is important that no alkali remains on the ground-glass joint.
[3] Recommended GLC conditions: ApL (3%) probably at a temperature between 120 and 150°.

EXPERIMENT 22. Trans-esterification

Both ester-formation from an acid and an alcohol and ester-hydrolysis occur under acid-catalysed conditions. The important step in these reactions is set out below, in simplified form, and careful examination will show that one is the reversal of the other.

ester-formation

$$\underset{H}{\overset{R'}{>}}O: \;\; \underbrace{\underset{R}{\overset{O}{\searrow}}C-\overset{+}{O}H_2}_{\text{acid*}} \;\rightleftharpoons\; \underbrace{\underset{H}{\overset{R'}{>}}\overset{+}{O}-C\underset{R}{\overset{O}{\lessgtr}}}_{\text{ester*}} + H_2O$$

ester-hydrolysis

$$\underset{H}{\overset{H}{>}}O: \;\; \underbrace{\underset{R}{\overset{O}{\searrow}}C-\overset{+}{O}\underset{H}{\overset{R'}{<}}}_{\text{ester*}} \;\rightleftharpoons\; \underbrace{\underset{H}{\overset{H}{>}}\overset{+}{O}-C\underset{R}{\overset{O}{\lessgtr}}}_{\text{acid*}} + R'OH$$

* shown in protonated form.

† An oven is not suitable for low-melting acids or for those, such as benzoic acid, which sublime readily.

What would happen if an ester ($R \cdot COOR'$) and an alcohol ($R''OH$) interacted in the presence of an acid catalyst?

$$R''(H)\ddot{O}: \; + \; O{=}C(R){-}\overset{+}{O}(H)R' \; \rightleftharpoons \; \text{——} \; + \; \text{——}$$

The result should be an equilibrium mixture of two alcohols and two esters. We can write the equation more simply as:

$$R \cdot COOR' + R''OH \rightleftharpoons R \cdot COOR'' + R'OH$$

The following experiment is designed to discover if this really happens. Methyl hexanoate is treated with varying amounts of methyl and propyl alcohols and the products are examined by GLC. To get meaningful results, information from several students should be brought together.

Place methyl hexanoate (13.0 g) in a round-bottomed flask (250 ml) along with the appropriate amount of methanol and propanol (see Table), concentrated sulphuric acid (1 ml)[1], and some anti-bumping granules. Fit the flask with a reflux condenser and heat on a steam bath for 45 minutes, by which time equilibrium should have been established.

*Ester**	*MeOH**†	*PrOH**	*Methyl:propyl ratio*	*Ester mixture Me(%)*	*Pr(%)*
13.0 g (0.10)	12.8 g (0.40)	30.0 g (0.50)	1:1	....	....
13.0 g (0.10)	7.4 g (0.23)	40.0 g (0.67)	1:2	....	....
13.0 g (0.10)	3.2 g (0.10)	48.0 g (0.80)	1:4	....	....
13.0 g (0.10)	–	54.0 g (0.90)	1:9	....	....

* These figures represent the weight and % mole of each component: the density of methanol is 0.79 and of propanol, 0.80.

† Methanol is added to the reaction mixture as well as propanol to give a similar reaction volume for each experiment.

Cool the reaction mixture and pour it into a separating funnel (250 ml) containing water (100 ml)[2]. Rinse the reaction flask with ether (2 x 50 ml) and add this to the separating funnel. Wash the ether solution with water (2 x 30 ml) and then dry the ether solution with a little anhydrous sodium sulphate. Filter the ether solution and distil off as much of the ether as possible from a steam bath[3]. The residual ester mixture containing a little ether is examined by GLC[4] to determine the proportions of the methyl and propyl esters. Insert your result into the above table and add results from other students.

Notes
[1] Add this carefully from a dropping-pipette to the swirled solution. *Do not suck the acid into the pipette.*
[2] Details for extraction and distillation are given briefly. Be sure to use apparatus of appropriate size. If additional instruction is required it can be adapted from details given in *Experiments 16 and 18.*
[3] Check your distillation apparatus with your instructor before using it for ether distillation.
[4] Recommended GLC conditions: PEGA (20%) at *ca.* 120°.

Questions. From the results you have collected do you think it would be possible to prepare a pure ester by a trans-esterification process? Can you suggest any way of disturbing the equilibrium to increase its value as a preparative procedure?

EXPERIMENT 23. Characterization of an unknown amine or phenol by acylation

Given an unknown amine or phenol, prepare a pure sample of the acetylated amine or the benzoylated phenol, determine its melting point, and identify the original compound as far as possible. For this experiment only one derivative need be prepared, but in identifying an unknown compound at least two points of identity are generally required.

EXPERIMENT 23a. Acetylation of the amine

Dissolve the amine (*ca.* 0.5 g) in glacial acetic acid (1 ml) contained in a small flask or tube (20 ml). Add acetic anhydride (1 ml), fit a condenser, and warm the mixture on the steam bath for 5 minutes. Pour the reaction mixture into cold water (10 ml) and stir well. A crystalline solid should separate; if it does not, carefully basify with strong ammonia solution. Filter off the precipitate and wash it with a little cold water. Crystallize the product from ethanol, aqueous ethanol, or water (try ethanol first). Determine its melting point and that of the amine if it is solid. Repeat the crystallization and melting-point determination.

Write an equation for the reaction you have carried out. Examine a list of melting points and report any which resemble the unknown.

	melting points	
	amine (if solid)	acetyl derivative
unknown amine		
possible amines (list names and give m.p.)		

EXPERIMENT 23b. Benzoylation of the phenol

Carry out this reaction in a small flask or tube (20 ml) with a well-fitting stopper (preferably ground glass).

To the phenol (*ca.* 0.5 g) add dilute sodium hydroxide (5 *M*, 8 ml) and benzoyl chloride (3 x 0.5 ml)[1,2]. After addition of each portion of benzoyl chloride, shake well for 2 or 3 minutes and cool under the tap or in ice-water if necessary. Finally shake for 5-10 minutes, when the smell of benzoyl chloride should have disappeared and the solution should still be alkaline (test with litmus paper). If benzoyl chloride is still present, add more alkali (1 ml) and shake for 5 minutes. Repeat this process if necessary.

Filter the solid which should separate at this stage, wash it with a little cold water, and crystallize from ethanol (or other appropriate solvent). Determine its melting point, and that of the phenol if it is solid. Repeat the crystallization and melting-point determination.

Write an equation for the reaction you have carried out. Examine a list of melting points and report any which resemble that of the unknown.

	melting points	
	phenol (if solid)	benzoate
unknown phenol		
possible phenols (list names and give m.p.)		

Notes [1] This is an unpleasant lachrymatory chemical and should be handled with care. Never try to wash it away with hot water. If necessary use concentrated ammonia solution to destroy it.

[2] The most common reason why this reaction fails is the use of too much benzoyl chloride. Use the quantities stipulated.

EXPERIMENT 24. Characterization of an unknown carboxylic acid as its amide or anilide

The conversion of a carboxylic acid into its amide or anilide is best achieved *via* the acid chloride, which can usually be obtained from the acid by interaction with thionyl chloride.

$$R\cdot COOH + SOCl_2 \longrightarrow R\cdot COCl \begin{cases} \xrightarrow{NH_3} R\cdot CONH_2 \quad \text{(amide)} \\ \xrightarrow{PhNH_2} R\cdot CONHPh \quad \text{(anilide)} \end{cases}$$

Convert the unknown acid into its acid chloride and thence to the amide or anilide. Determine its melting point and identify the acid as far as possible. For this experiment only one derivative need be prepared, but in identifying an unknown compound at least two points of identity are generally required.

EXPERIMENT 24a. Preparation of the acid chloride

Place the acid (*ca.* 0.5 g) in a small flask along with thionyl chloride (5 ml), fit a reflux condenser, and boil gently for 30 minutes with a small Bunsen burner in the fume cupboard. Remove the excess of thionyl chloride by connecting the reaction flask to the water pump and swirling the flask while keeping it warm on a steam bath[1]. These conditions may have to be moderated if the acid chloride is volatile.

EXPERIMENT 24b. Preparation of an amide

This reaction may be vigorous: operate carefully in a fume cupboard. Place concentrated ammonia solution (*ca.* 10 ml) in a conical flask and slowly add the acid chloride to it. Rinse out the flask containing the acid chloride with a little ammonia solution (*ca.* 2 ml) and add this to the main product. Cool the solution, filter off the precipitate, wash it with a little water, and crystallize it from ethanol, aqueous ethanol, or water (try ethanol first). Determine the melting point of the amide, and of the acid if this is solid. For details of how to report your results see after *Experiment 24c.*

EXPERIMENT 24c. Preparation of an anilide

This reaction may be vigorous: operate carefully in a fume cupboard. Add aniline (1 ml) dropwise to the acid chloride with frequent swirling. After a few minutes, add dilute hydrochloric acid (10 ml) to dissolve the excess of aniline and stir thoroughly to break any lumps of solid which are present. Filter the crude anilide, wash it with cold water, and crystallize it from ethanol. Determine the melting point of the anilide, and of the acid if this is solid.

Write an equation for the reaction you have carried out. Examine a list of melting points and report any which resemble that of the unknown.

	melting points	
	acid (if solid)	amide/anilide
unknown acid		
possible acids (give names and m. p.)		

Note [1] Consult your instructor if you are not sure how to do this.

6. Detection of amines (and alcohols)

Objectives

(1) To illustrate the value of N—H and O—H stretching bands as a means of detecting amines and alcohols.
(2) To study the reaction of different classes of amines with nitrous acid.

Introduction

An organic amine will normally have the following characteristics:

(i) It can be shown to contain nitrogen by the sodium fusion test (p. 145).
(ii) It will be basic and this will usually be apparent from its relative solubility in water and in dilute acid. This difference is most easily observed when the compound appears to be insoluble in cold water and soluble in dilute aqueous acid.
(iii) It shows characteristic behaviour with nitrous acid.
(iv) Primary amines (RNH_2), secondary amines (R_2NH), and other compounds containing an N—H bond have an infrared absorption band of variable intensity in the region 3150-3500 cm^{-1}. (Some hydroxy compounds also absorb in this region.)

This exercise will deal with (iii) and (iv) only.

EXPERIMENT 25. The infrared spectra of amines (and alcohols)

Compounds containing C—H, N—H, and O—H groups have absorption bands in the frequency ranges set out below. These are discussed in more detail on pp. 28-29.

C—H	3100-2850 cm^{-1}
N—H	3500-3150 cm^{-1}
O—H	3600-3200 cm^{-1}

Hydrogen bonded OH groups absorb at the lower end of this frequency range. Free OH groups, which are likely to be encountered only in dilute solution, absorb in the range 3650-3590 cm^{-1}.

Mark these three regions distinctively on the infrared chart already used in Experiment 13.

Draw up a table like that below and examine the spectra of acetanilide, aniline, benzaldehyde, benzamide, benzoic acid, butan-2-ol, cyclohexylamine, diethyl ether, *N,N*-dimethylaniline, ethanol, ethyl acetate, ethyl acetoacetate, and *N*-methylaniline. Insert the structures of the compounds listed, mark any O—H, N—H, or C=O bonds in the structures, and insert the frequencies of the appropriate absorption bands from the spectra. When you have completed the table, check it with your instructor.

Name	*Structure*	*N–H or O–H band*	*C = O band*
Acetanilide			
Aniline			
etc.			

EXPERIMENT 26. The reactions of amines with nitrous acid

Amines can be classified as aliphatic or aromatic and as primary, secondary, or tertiary, making *six* classes in all. It is sometimes useful to distinguish between these six classes, and their reaction with nitrous acid goes some way toward this.

An aromatic amine has its nitrogen atom directly attached to at least one aryl group; an aliphatic amine has its nitrogen atom attached to alkyl groups or hydrogen only. An amine such as $C_6H_5 \cdot CH_2NH_2$ with the NH_2 group attached to an aliphatic carbon atom will behave as an aliphatic amine, even though it contains an aromatic ring.

Primary (RNH_2), secondary (R_2NH), and tertiary amines (R_3N) are distinguished by the number of alkyl and/or aryl groups attached directly to nitrogen.

Classify each compound formulated below as an *aliphatic amine,* an *aromatic amine* or *not an amine.* Classify the amines as primary or secondary or tertiary.

$EtNH_2$, $PhNHMe$, Et_2NH, $PhNEt_2$, $PhCH_2NHMe$,

$PhCONH_2$, Me_2NEt, Me_2NH, $PhNH_2$, $EtCONHMe$

With nitrous acid the six types of amines react as follows:

	Aliphatic	*Aromatic*
Primary	Nitrogen liberated, alcohol (and other compounds) formed.	Diazonium salt formed, gives azo dye with phenols
Secondary	Insoluble *N*-nitrosoamine formed	Insoluble *N*-nitrosoamine formed
Tertiary	No visible reaction	Gives *C*-nitroso compound, usually green

Carry out the following test with the two amines provided by your instructor and classify them as far as you can. If you have difficulties with these tests, try them with known compounds to observe the expected reactions.

Dissolve the amine (*ca.* 0.2 g) in dilute hydrochloric acid (2 *M*, 3 ml) and cool the solution to 0-5° in an ice bath. Prepare a solution of sodium nitrite (*ca.* 0.4-0.5 g) in water (1 ml) and add this dropwise to the amine solution which should be well stirred or shaken. It is important that the reaction temperature should not exceed 5° throughout. Watch out for three possible results:

(i) Steady evolution of nitrogen indicates the presence of a **primary aliphatic amine.** Take care not to confuse this observation with slow evolution of gas through decomposition of nitrous acid.

(ii) The appearance of a precipitate (usually a liquid but sometimes a solid) indicates a **secondary amine.** Both aliphatic and aromatic amines give the same reaction. Since aromatic tertiary amines sometimes give a brown oil or precipitate at this stage, basify the solution. If a tertiary aromatic amine is present, the solution will turn green and a precipitate (solid or liquid) should appear.

(iii) If the reactions described in (i) and/or (ii) are not observed, divide the solution into two parts and test each portion:

(*a*) Add one portion to a solution of β-naphthol (0.1-0.2 g) in dilute sodium hydroxide (1 ml). An orange or red precipitate will be formed if a primary aromatic amine is present.

(*b*) If test (*a*) is negative add alkali to the remaining portion. If a tertiary aromatic amine is present, the solution will turn green and a precipitate (solid or liquid) should appear.

Report your conclusions, state your reasons briefly, and give equations for the reactions observed. Consult a textbook if this is necessary.

7. The identification of unknown organic compounds

Objective

To provide an introduction to the procedures, both spectroscopic and chemical, by which organic compounds are identified.

Introduction

The identification of an unknown compound already known to chemists involves a series of investigations which fall into two stages. (If the substance has never been identified before, the approach has to be modified somewhat.)

(i) The recognition of the functional groups present. This is achieved in research laboratories by a range of spectroscopic, chromatographic, physical, and chemical studies. Simple procedures appropriate at this stage are outlined below.

(ii) The identification of the substance by proving its identity with some known compound. This is achieved by comparing spectroscopic, chromatographic, and physical and chemical properties of the unknown with those of the substance with which it is believed to be identical.

This exercise is designed to provide an introduction to this important aspect of practical organic chemistry. A fuller account of this topic is given on pp. 143-166. Experiments 27, 28 and 29 should be carried out in that order. Check with your instructor at each stage that you are qualified to proceed further.

EXPERIMENT 27. Detection of acidic and basic functional groups

Read Section 6 (p. 147).

Acidic compounds usually contain a sulphonic acid (SO_3H), phenol (aromatic OH), or carboxylic acid (COOH) group. The most common **basic substances** are amines (primary, secondary and tertiary). Assuming that the following substances are insoluble in water, designate them as *soluble* or *insoluble* in dilute acid and dilute alkali (R = alkyl group, Ar = phenyl or other aryl group). Check these with your instructor before carrying out the tests that follow.

$$RCOOH, RCH_2OH, RSO_3H, RCOR', R_2NH, RCH(NH_2)CO_2H$$

$$ArNH_2, ArCOOH, ArOH, ArSO_3H, ArCONH_2, HO_3SC_6H_4CH_2NH_2$$

Carry out Test 6 (p. 147) on substances **A**–**E**. Report your results in a table with the following headings:

	Substance	*Water*	*Acid*	*Alkali*	*Conclusion*
e.g.	**Z**	insoluble	insoluble	soluble	an acidic compound

EXPERIMENT 28. Lassaigne's (sodium fusion) test (Test 4, p. 145)

Carry out this test on substance **F**. It is essential to carry out this test correctly and if necessary it should be repeated on other substances until satisfactory results are consistently obtained.

EXPERIMENT 29. Identification of an unknown compound

Carry out this identification in the following stages checking with your instructor at each stage.

(i) Carry out Lassaigne's (sodium fusion) test on substance **G** and examine the infrared spectrum provided, particulary for N—H, O—H, and C=O groups. Draw what conclusions you can about functional groups which may be present.

(ii) Discuss with the instructor what additional tests would help to confirm the conclusions already reached and to gain additional information about **G**. Carry out these tests.

(iii) Consider what derivative would be suitable for the further characterization of **G**. Check with your instructor, then prepare the derivative using 0.5 g of unknown compound. Confirm your conclusion by melting point, mixed melting point if possible, or any other appropriate means. If **G** is a solid, use its melting point as additional information.

For this experiment only one derivative need be prepared, but in identifying an unknown compound at least two points of identity are generally required.

Report your results under the following headings:

(i) Result of Lassaigne's (sodium fusion) test.
(ii) Conclusion (if any) from infrared spectrum.
(iii) Brief account of additional tests.
(iv) Derivative: name this and record its melting point; there is no need to describe its preparation.
(v) Conclusion: indicate any compound(s) which may be identical with **G** and quote appropriate melting points for comparison.

8. Electrophilic aromatic substitution (nitration)

Objectives

(1) To compare (*a*) the reaction conditions and (*b*) the products, in the nitration of nitrobenzene, benzaldehyde, bromobenzene, and acetanilide.
(2) To illustrate some techniques used for the detection of by-products.

Introduction

Substitution reactions of aliphatic compounds are predominantly reactions with nucleophiles (see p. 77), but aromatic compounds undergo substitution mainly with electrophiles such as the nitronium ion (NO_2^+, this section) or the bromonium ion (Br^+, p. 106). Reaction is known to occur through a resonance-stabilized carbonium ion.

$$C_6H_6 + X^+ \rightleftharpoons [C_6H_6X]^+ \rightleftharpoons C_6H_5X + H^+$$

With monosubstituted benzene compounds (C_6H_5Z) the additional group (Z) controls both the reactivity of the system (and hence the conditions of reaction) and the proportion of isomeric products.

The nitration of benzene is effected smoothly with a mixture of concentrated nitric acid and concentrated sulphuric acid at a temperature between 20 and 60°. The two acids interact to produce the nitronium ion which is the reactive species effecting nitration.

It is suggested that students carry out one of the following nitration reactions and then with the help of other students complete the Table given after Experiment 33.

EXPERIMENT 30. Nitration of nitrobenzene

Carefully add *fuming* nitric acid (*d* 1.5, 5 ml) to ice-cold concentrated sulphuric acid (7 ml) in a round-bottomed flask (100 ml), and carefully add nitrobenzene (4 g, 3.3 ml) portionwise and with swirling, to the mixed acid. Attach an air condenser to the flask and heat on the steam bath for one hour. Pour the reaction mixture into ice-water (*ca.* 80 ml), filter off the solid, wash it well with water, and recrystallize from methanol. Collect a second crop of crystals after concentrating the methanol filtrate. Report the results as indicated in the Table (p. 105) and write an equation for the reaction.

EXPERIMENT 31. Nitration of benzaldehyde

Carefully add *fuming* nitric acid (*d* 1.5, 4.5 ml) to ice-cold concentrated sulphuric acid (31 ml) in a conical flask (250 ml). Cool this mixture in an ice-bath and add freshly distilled benzaldehyde (5.3 g, 5.1 ml) *dropwise*, with swirling, at such a rate that the temperature of the reaction mixture does not exceed 10°. Remove the ice-bath and allow the mixture to stand for one hour before pouring it on to crushed ice (*ca.* 150 g). Filter off the product; wash it well with water, sodium bicarbonate solution (keep these alkaline washings), and again with water. Recrystallize the product from a mixture of ethanol and water.

Carefully acidify the bicarbonate washings. If a precipitate forms, filter it off, wash it with water, and recrystallize it from boiling water. Try to work out what this alkali-soluble by-product might be and check your conclusions in some suitable way (mixed melting point, infrared spectrum).

Report the results as indicated in the Table (p. 105) and write an equation for the reaction.

EXPERIMENT 32. Nitration of bromobenzene

Add concentrated sulphuric acid (10 ml) *slowly*, with cooling and swirling, to concentrated nitric acid (*d* 1.42, 10 ml) in a round-bottomed flask (100 ml). Slowly add bromobenzene (8 g, 5.3 ml) in portions of *ca.* 1 ml, swirling the mixture during addition, and allowing the temperature to rise to 50-60° but *not higher* (cool in ice if necessary). When the addition is complete, and the temperature no longer tends to rise, fit a water-condenser to the flask, and heat the mixture at 100° for 30 minutes, shaking the flask from time to time. Cool the mixture and add it to ice-water (100 ml). Filter off the solid, wash it with water, and recrystallize it from ethanol. The first crop of crystals is an almost pure isomer of bromonitrobenzene. Identify this by its melting point and, if possible, mixed melting point.

Another isomer is more soluble in ethanol and remains in the mother liquor. Evaporate the ethanolic filtrate to dryness, weigh the residue, and dissolve it in a little ether. Examine it by GLC using an ApL column (10% at 170°), and try to identify the peaks by comparison of the retention times with those for authentic samples of bromobenzene and (if available) the three bromonitrobenzenes.

Enter the results into the Table (p. 105) and write an equation for the reaction.

EXPERIMENT 33. Nitration of acetanilide

Add concentrated sulphuric acid (10 ml) slowly, with cooling and swirling, to a suspension of acetanilide (5 g) in acetic acid (5 ml) contained in a conical flask (100 ml). Cool the mixture in an ice-bath, and add a mixture of concentrated nitric acid (2.2 ml) and concentrated sulphuric acid (1.4 ml) *dropwise*, with swirling, at such a rate that the temperature does not exceed 10°. Then remove the ice-bath, let the mixture stand for one hour, and pour it on to crushed ice. Filter off the solid product, wash it with water, and recrystallize the nitroacetanilide from ethanol. A second nitroacetanilide isomer remains in the ethanol mother liquor and should be isolated as follows.

Evaporate the ethanolic filtrate to dryness and extract the residue with a mixture (1:1, 50 ml) of benzene and ether. Filter off the insoluble and almost colourless solid. The solution remaining may contain mono- and di-nitroacetanilides, unreacted acetanilide, and nitroanilines. It is chromatographed on a column of silica gel (*ca.* 50 g) made up in benzene (see p. 15). Develop the column with ether-benzene (1:1) mixture, and collect the pale yellow eluate. Distil off the solvent and recrystallize the residue from aqueous methanol.

Identify the products by comparison (m.p., infrared spectrum) with authentic samples.

Enter the results into the Table opposite and write an equation for the reaction.

Results and Questions

Tabulate results obtained by several students as shown below and answer the questions that follow:

Starting compound	*Reaction conditions*			*Principal product*	*Yield g (%)*	*m.p. found (lit)*	*By-products*
	HNO_3	H_2SO_4	*Temp.*				
Nitrobenzene	Fuming	Conc.	100°	?-dinitro-benzene			
Benzaldehyde							
Bromobenzene							
Acetanilide							

(1) In each case comment on the reaction conditions, the isomer(s) formed, by-products (if any), and any other significant features of the results.

(2) Suggest reaction conditions for the mononitration of anisole, benzoic acid, phenol, and toluene. What products result from these reactions?

(3) Why is the sequence given below used for the preparation of *o*- and *p*-nitroaniline in preference to the direct nitration of aniline?

$$C_6H_5NH_2 \xrightarrow{(CH_3CO)_2O} C_6H_5NHCO\cdot CH_3 \xrightarrow[H_2SO_4]{HNO_3,}$$

$$o/p\text{-}O_2NC_6H_4NHCO\cdot CH_3 \xrightarrow{H_2O,\ H^+} o/p\text{-}O_2NC_6H_4NH_2$$

(4) Suggest what products would be likely to result from the nitration of benzaldehyde if the reaction were carried out at 100° instead of 0-10°.

9. Electrophilic aromatic substitution (bromination)

Objectives

(1) To compare the efficiency of four different catalysts (iron wire, iron filings, pyridine, and silver nitrate) in the bromination of benzene.
(2) To provide further experience in the technique of fractional distillation and in the use of GLC to examine product purity.

Introduction

Read the introduction on p. 102. It is suggested that each student carries out one of the following bromination experiments. The Table on p. 108 should then be completed with the help of other students.

EXPERIMENT 34. Bromination in the presence of iron wire

Add bromine (5 ml)[1] * to benzene (25 ml) in a round-bottomed flask (100 ml). Add clean iron wire (0.15 g) and at once fit a reflux condenser to the flask[2] *. A vigorous reaction soon sets in; hydrogen bromide is evolved and the mixture boils. Moderate the reaction by cooling the flask in a cold-water bath. When spontaneous refluxing ceases, heat the solution under reflux on the steam bath until the *condensate* no longer contains bromine (i.e. becomes colourless). Cool the reaction mixture, and decant the liquid into a separating funnel containing water (15 ml). Shake and separate the layers; dry the organic layer[3] * over calcium chloride or magnesium sulphate.

Examine the product as directed on p. 107.

EXPERIMENT 35. Bromination in the presence of iron filings

Carry out the bromination described in *Experiment 34* replacing iron wire by an equal weight of iron filings[2] *.

Examine the product as directed on p. 107.

* Notes are given on p. 108.

EXPERIMENT 36. Bromination in the presence of pyridine

Add dry pyridine (0.1 ml) to benzene (25 ml) in a round-bottomed flask (100 ml). Attach a reflux condenser to the flask[2]*, and carefully add bromine (5 ml,[1]*) down the condenser. A vigorous reaction soon occurs, with the evolution of hydrogen bromide; cool the flask in a cold-water bath if the reaction becomes too vigorous. When the reaction subsides, warm the mixture to 40° for about one hour and then heat under reflux for a further hour. Cool the reaction mixture, pour it into a separating funnel, wash it with water[3]*, dilute sodium hydroxide (2 *M*), and again with water, and dry it over calcium chloride or magnesium sulphate.

Examine the product as directed below.

EXPERIMENT 37. Bromination in the presence of silver nitrate

Stir benzene (10 ml), bromine (2 ml,[1]*), and nitric acid (2 *M*, 350 ml) in a conical flask (1 litre) wrapped in metal foil. Add a solution of silver nitrate (8 g) in water (35 ml) gradually over one hour; then remove the metal foil and add sodium sulphite in small portions until the bromine colour is discharged. Extract the product with ether or methylene chloride (50 ml), wash the extract[3]* with water, dry it over calcium chloride or magnesium sulphate, and distil off the solvent on a steam bath.

Examine the product as directed below.

Separation and examination of the bromination products

Transfer the product to a Claisen flask (25 ml) and distil the mixture slowly from an oil bath. Collect three fractions: (1) b.p. below 100° (mainly unreacted benzene), (2) 100-140°, and (3) 140-165°. Record the weight and exact boiling range of each fraction.

Use a hot dropping-pipette to transfer the residue from the distillation flask to a small weighed specimen tube and stopper this immediately[4]*. This fraction will probably solidify on cooling. It can be purified by crystallization from methanol or by elution from an alumina column with dry petroleum (b.p. 40-60°).

Examine each distilled fraction and the residue (before purification) by GLC (10% ApL, 140°). It will be necessary to make separate injections of benzene, bromobenzene, and *o*- and *p*-dibromobenzene for comparison purposes.

* Notes are given on p. 108.

Pure bromobenzene can be obtained by redistillation of fraction 3 (along with fraction 2 if this contains much bromobenzene). Collect the fraction boiling between 154 and 157°.

Notes [1] Bromine must be handled with great care in the fume cupboard. The liquid produces painful burns on the skin, and the vapour is toxic and extremely irritant. See also the footnote on p. 67‡.
[2] Hydrogen bromide is evolved in these experiments which should therefore be conducted in a fume cupboard.
[3] Be careful to preserve the correct layer. If in doubt, test a few drops of each layer for solubility in water.
[4] The dibromobenzene is a volatile solid which will evaporate rapidly at room temperature if left in an open vessel.

Results

Catalyst	*Bromobenzene g (%)*	*b.p.*	*Dibromobenzene isomer(s) g (%)*	*m.p.*	*Other products (if any)*
Iron wire Iron filings Pyridine Silver nitrate					

Comment on any differences in yields obtained by the different methods. Write the mechanism for the mono-bromination of benzene with an iron catalyst and suggest what part the pyridine and silver nitrate play in catalysing the reaction.

10. The formation and reactions of diazonium salts

Objectives

(1) To provide experience in the preparation and use of diazonium salts.
(2) To illustrate the value of column chromatography as a method of purification.
(3) To offer practice in the adaptation of experimental instructions to different compounds of the same class.

Introduction

Compared with their aliphatic counterparts, aromatic primary amines give relatively stable diazonium salts by reaction with nitrous acid. Although dangerously explosive when dry, these salts are moderately stable in aqueous solution at 0° and in this form they are versatile reagents in organic synthesis.

As illustration of this last point, list the reagents required to convert the diazonium salt $Ar\overset{+}{N}_2\ \bar{X}$ into each of the following (consult a textbook where necessary):

$$ArOH, ArBr, ArCN, ArH, ArPh, ArF, ArN{=}NNHAr', ArN{=}NC_6H_4NH_2(p)$$

Reactions of diazonium salts often give several by-products including phenols, azo compounds, and diazoamino compounds in addition (sometimes) to unreacted amine. In these experiments purification is effected by column chromatography.

Suggest a reason why phenols, azo compounds and diazoamino compounds may be formed as by-products. What product would you expect from the interaction of *o*-diaminobenzene with one mole of nitrous acid?

It is expected that the reactions in this section will be carried out, not with aniline, but with some other aromatic amine. Toluidine (*o, m,* and *p*), anisidine (*o* and *p*), chloroaniline (*o* and *p*), *p*-bromoaniline, and 2,4- and 2,5-dichloroaniline give satisfactory results. The weakly basic nitroanilines and 2,4,6-tribromoaniline are not suitable for these instructions. The directions given will have to be modified in respect of the amounts of reagents used. In Experiments 40 and 41 the product should be purified by chromatography followed, as appropriate, by crystallization or distillation (possibly under reduced pressure). If it is obvious that the starting amine is very impure (i.e. very dark in colour), it should be distilled or crystallized before use.

EXPERIMENT 38. Diazotization of aniline

Shake a mixture of aniline (5 g), concentrated hydrochloric acid (15 ml), and water (15 ml) in a conical flask (250 ml) until the solution is clear, warming if necessary. Cool the solution in an ice-water mixture, swirling it so that, if aniline hydrochloride separates, it does so in the form of small crystals.

Whilst still swirling add *dropwise* a solution of sodium nitrite (4 g) in water (10 ml) at such a rate that the temperature does not rise above 10°. When the addition is complete, the solution should be clear; if a precipitate remains, filter the solution through a small plug of glass wool into another ice-cold flask.

Keep the solution cold and use it for *Experiment 39* and for *Experiment 40* or *41*.

EXPERIMENT 39. Diazo coupling

Slowly add the diazonium salt solution (1 ml) to a well-shaken solution of β-naphthol (0.3 g) in 2 *M* sodium hydroxide solution (3 ml). Filter off the red precipitate, recrystallize it from acetic acid, and record its m.p.. Write an equation for this reaction.

EXPERIMENT 40. Preparation of chlorobenzene (Sandmeyer reaction)

Add the remaining benzenediazonium chloride solution (see *Experiments 38* and *39*) very slowly to a mechanically stirred ice-cold solution of cuprous chloride (5.5 g) in concentrated hydrochloric acid (20 ml) contained in a conical flask (250 ml,[1]). When the reaction subsides, warm the mixture gently to about 70-80° on a water bath until no more nitrogen is evolved. Then transfer the mixture to a round-bottomed flask (250 ml), add water until the flask is half-full, and distil the mixture[2] until the distillate becomes clear.

Extract the distillate with ether and dry the extract with magnesium sulphate. Distil off the ether and dissolve the residue in the minimum volume of dry petroleum (b.p. 40-60°).

Prepare a column of alumina (*ca.* 100 g) in petroleum (b.p. 40-60°) and chromatograph the product, using more petroleum as eluant (see p. 15). The chlorobenzene is not strongly adsorbed and should be eluted by 100-200 ml of petroleum. Collect the colourless eluate, distil off the solvent, and distil the residual chlorobenzene from an oil bath.

Write an equation for the reaction and consult a textbook for the mechanism. This involves one-electron transfers in which copper acts first as a reducing agent and then as an oxidizing agent.

Notes [1] The large flask is necessary because nitrogen evolution is accompanied by frothing. Stirring with a glass rod may also be necessary to break up the froth.

[2] This is a simple form of steam distillation.

EXPERIMENT 41. Preparation of iodobenzene

Add a solution of potassium iodide (9 g) in water (15 ml) *slowly* and with mechanical stirring to the remainder of the diazonium salt solution (see *Experiments 38* and *39*). Additional stirring with a glass rod may be necessary to break up the froth. Nitrogen is evolved. Allow the mixture to stand for one hour, then heat it on the steam bath until nitrogen evolution ceases. Transfer the mixture to a round-bottomed flask (250 ml), add water until the flask is about half-full, and distil the mixture until the distillate becomes clear [1].

Extract the distillate with ether; wash the extract with sodium bisulphite solution (to remove free iodine) and water. Dry the ether solution over magnesium sulphate and distil off the solvent. Chromatograph the product as described for chlorobenzene and finally distil the iodobenzene under reduced pressure.

Write an equation for the reaction.

Note [1] This is a simple form of steam distillation.

11. The formation and reactions of Grignard reagents

Objectives

(1) To illustrate the preparation and usefulness of Grignard reagents.
(2) To obtain experience of reactions carried out under anhydrous conditions.

Introduction

Grignard reagents, like diazonium salts (p. 109), are reactive intermediates which are widely used in synthesis. They resemble diazonium salts also in that they are dangerously explosive when dry, and so are always prepared and used in solution.

The preparation of a Grignard reagent demands absolutely anhydrous conditions. *Both the apparatus and the reagents must be scrupulously dried before the preparation is attempted.* Grignard reagents react very readily with water.

The Grignard reagent ($\overset{\delta-}{R}-\overset{\delta+}{Mg}X$) behaves as a nucleophile. Formulate the product resulting from interaction of RMgX with an aldehyde (R′CHO), ketone (R'_2CO), ester ($R'CO_2Me$), and with water, carbon dioxide, and ethylene oxide. The reactions with aldehydes, ketones, and esters are illustrated in the following experiments.

EXPERIMENT 42. Preparation of phenylmagnesium bromide

Fit a three-necked flask (25 ml) with a dropping funnel and condenser; fit both of these with calcium chloride tubes. Add dry magnesium turnings (0.4 g,[1]) to the flask and add, from the dropping funnel, 2-3 ml of a solution of bromobenzene (2.65 g) in dry ether (9 ml,[1]).

The formation of the Grignard reagent is indicated by the appearance of a cloudiness in the solution and the production of heat. If the reaction does not start, grind the magnesium pieces against each other with a glass rod. The ether should boil gently.

When the initial reaction subsides (**not before**) add the remainder of the bromobenzene solution at such a rate that the ether boils gently. Rinse out the funnel with dry ether and add the rinsings to the flask. Then boil the mixture on a steam bath for 15 minutes.

Use this Grignard reagent immediately for experiment 43 or 44.

Note [1] The magnesium should be heated in the oven for at least one hour and cooled in a desiccator. The ether should be sodium-dried; if this is not available consult your instructor.

EXPERIMENT 43. Preparation of diphenylmethanol (benzhydrol)

Add to the Grignard reagent (*Experiment 42*) a solution of redistilled benzaldehyde (1.6 g) in dry ether (4 ml). Boil the mixture for 15 minutes, cool, and carefully pour on to crushed ice (5-10 g). Shake the mixture with dilute hydrochloric acid (3 ml), separate the ether layer, and wash it with water[1]*, sodium bisulphite solution, and again with water. Distil off the ether, add water to the residue and distil this mixture until the distillate is clear. The *non-volatile residue*[2]* contains the reaction product. Cool the residue, filter off the solid, and recrystallize from petroleum (b.p. 40-60°).

If possible, recover the volatile products by extraction of the aqueous distillate and try to identify them by GLC comparison with authentic substances (unreacted aldehyde, bromobenzene, biphenyl).

Write a mechanism for this reaction and suggest how biphenyl might be formed.

EXPERIMENT 44. Preparation of triphenylmethanol

Add to the Grignard reagent (*Experiment 42*) a solution of benzophenone (2.6 g) or methyl or ethyl benzoate (1.2 g) in dry ether (4 ml), and boil the mixture for 15 minutes. Cool the contents of the flask and pour on to crushed ice (5-10 g). Shake the mixture with dilute hydrochloric acid (5 ml), separate the ether layer, and wash it well with water[1]*. Distil off the ether, add water to the residue, and steam distil to remove impurities. Cool the *non-volatile* residue [2]*, filter off the triphenylmethanol, and recrystallize it from methanol.

If possible, recover the volatile products by extraction of the aqueous distillate and try to identify them by GLC comparison with authentic substances (unreacted ketone or ester, bromobenzene, biphenyl).

* Notes are given on p. 114.

Write a mechanism for this reaction and suggest how biphenyl might be formed.

Notes

[1] In carrying out this extraction it may be necessary to use additional ether. It is important to wash out all the mineral acid from the ether extract since this may catalyse the conversion of the alcohol to an ether *via* a resonance-stabilized carbonium ion. Bisdiphenylmethyl ether (m.p. 110°) and methyl triphenylmethyl ether (two forms, m.p. 83 and 96°) are sometimes obtained from Experiment 43 and 44 respectively.

[2] In steam distillation the distillate is commonly required, but in this experiment unwanted volatile products are being separated. *The product remains in the distillation flask and this must not be allowed to boil dry.*

12. Selective reduction of carbonyl compounds with metal hydrides

Objectives

(1) To illustrate the use of sodium borohydride and lithium aluminium hydride as selective reducing agents.
(2) To obtain further experience of carrying out reactions under anhydrous conditions.

Introduction

Metal hydrides are valuable reagents for the selective reduction of various functional groups. Reduction occurs in two stages and requires hydride ion (H^-) for the first step and proton (H^+) for the second,

$$\text{e.g.}\quad R\cdot \overset{\delta^+}{C}H{=}\overset{\delta^-}{O} \;\; H{-}\bar{A}lH_3\,Li^+ \longrightarrow R\cdot CH_2O\bar{A}lH_3\,Li^+ \xrightarrow{3R\cdot CHO} (R\cdot CH_2O)_4\bar{A}l\,Li^+ \xrightarrow{H_2O} 4R\cdot CH_2OH + Al(OH)_3 + LiOH$$

Lithium aluminium hydride ($LiAlH_4$) reduces all carbonyl compounds (including carboxylic acids and their derivatives), oximes, and nitro compounds. Sodium borohydride is a weaker reducing reagent reacting only with aldehydes, ketones, and acid chlorides. Insert the structures of the products of the following reactions or write 'no reaction'. Discuss your answers with your instructor.

Product	$NaBH_4$	Substrate	$LiAlH_4$	Product
———	$\longleftarrow$	$R\cdot COCl$	$\longrightarrow$	———
———	$\longleftarrow$	$R\cdot CHO$	$\longrightarrow$	———
———	$\longleftarrow$	$R\cdot COOH$	$\longrightarrow$	———
———	$\longleftarrow$	$R\cdot CO_2R'$	$\longrightarrow$	———
———	$\longleftarrow$	$R\cdot CO\cdot R'$	$\longrightarrow$	———
———	$\longleftarrow$	$R\cdot CH_2NO_2$	$\longrightarrow$	———
———	$\longleftarrow$	$R\cdot CH{=}NOH$	$\longrightarrow$	———
———	$\longleftarrow$	$R\cdot CONH_2$	$\longrightarrow$	———

Neither of these reagents reduces isolated olefinic (C=C) or acetylenic (C≡C) bonds, but lithium aluminium hydride sometimes reduces them when they are conjugated with a carbonyl or hydroxyl group.

$$Ph \cdot CH{=}CH \cdot CHO \begin{cases} \xrightarrow{NaBH_4} \text{———} & \text{(reduction of carbonyl group only)} \\ \xrightarrow{LiAlH_4} \text{———} & \text{(reduction of both unsaturated centres)} \end{cases}$$

EXPERIMENT 45. Reduction of diphenylacetic acid with lithium aluminium hydride

$$\underset{\text{diphenylacetic acid}}{Ph_2CH \cdot COOH} \xrightarrow{LiAlH_4} \underset{\text{2,2-diphenylethanol}}{Ph_2CH \cdot CH_2OH}$$

Lithium aluminium hydride is caustic and reacts violently with water giving hydrogen which may ignite. All apparatus must be scrupulously dry.

Weigh lithium aluminium hydride (0.2 g) into a round-bottomed flask (50 ml)[1], add ether (10 ml) which has been dried over sodium wire, and fit a reflux condenser to the flask.

Dissolve diphenylacetic acid (0.6 g) in dry ether (*ca.* 10 ml) and add this solution dropwise down the condenser, using a dropping pipette and swirling the flask during the addition. This should take about 10 minutes.

Fill a drying tube with calcium chloride and attach it to the top of the condenser. Reflux the reaction mixture for one hour on a steam bath, swirling occasionally, then cool the flask and decompose the excess of hydride by *slowly* adding 'wet' ether (15 ml)[2] down the condenser. Reflux for 10 minutes, cool, and *slowly* add dilute sulphuric acid (10 ml) to liberate the alcohol from its alkoxide[3]. Separate the ether layer with a separating funnel, wash with water (10 ml), and dry the ether solution over anhydrous sodium sulphate. Filter, and distil off the volatile solvent on a water bath. Cool the residue in ice until it solidifies. If it fails to solidify, it probably contains residual ether which must be removed under reduced pressure. Crystallize the product from petroleum (b.p. 60-80°). If possible, record the infrared spectra of the starting material and the product and notice the disappearance of the carbonyl absorption band and the appearance of the O–H stretching band. Write a brief account of your experiment and report the weight and melting point of your product.

2,2-diphenylethanol
wt. . . . . . . g (. . . . . . %), m.p. . . . . . . (lit. . . . . . .).

Notes [1] Your instructor may wish to assist you with this operation.
[2] 'Wet' ether (containing *ca.* 7% water) is prepared by shaking ether with water. Use only the upper layer. Ethyl acetate, which is normally used for this purpose, leads to the formation of 2,2-diphenylethyl acetate.
[3] If a precipitate persists at this stage, add more dilute sulphuric acid.

EXPERIMENT 46. Reduction of diphenyl ketone (benzophenone) with sodium borohydride

$$Ph_2C{=}O \xrightarrow{NaBH_4} Ph_2CHOH$$

The less reactive sodium borohydride can be used in aqueous or alcoholic solution without serious loss of reagent [1].

Dissolve benzophenone (0.75 g) in methanol (10 ml) in a round-bottomed flask (25 ml). Add sodium borohydride (0.2 g) carefully, and allow the mixture to stand at room temperature with occasional swirling for 20 minutes. Distil off the bulk of the methanol on a steam bath, cool, and pour the residue into a separating funnel containing water (20 ml). Rinse out the reaction flask with ether (2 x 10 ml) and add this to the separating funnel. Dry the ethereal extract over sodium sulphate or magnesium sulphate. After removal of the ether on a steam bath crystallize the residue from petroleum (b.p. 60-80°). If possible, record the infrared spectra of the starting material and the product, and notice the disappearance of the carbonyl absorption band and the appearance of the O—H stretching band. Write a brief account of your experiment and report the weight and melting point of your product.

diphenylmethanol (benzhydrol)
wt. . . . . . . g (. . . . . . %), m.p. . . . . . . (lit. . . . . . .)

Notes [1] Decomposition of the reducing agent can be minimized by carrying out the reaction in methanolic sodium hydroxide (0.2 *M*), but this has little effect on the yield of product.

13. The condensation of carbonyl compounds with carbanions

Objectives

(1) To illustrate one of the many types of base-catalysed condensation reactions between carbonyl compounds and potential carbanions.
(2) To study the ultraviolet spectra of unsaturated carbonyl compounds.

Introduction

Many reactions involve the condensation of a carbonyl compound with a second compound which is said to contain a reactive methylene group. Such reactions are frequently base-catalysed and the main purpose of the base is to generate a carbanion (usually resonance-stabilized) which then reacts with the carbonyl compound. At this stage a β-hydroxycarbonyl compound is usually formed and this commonly eliminates water to furnish an αβ-unsaturated carbonyl compound.

The Knoevenagel reaction is a typical condensation of this type in which

$$R\cdot CHO + CH_2(CO_2H)_2 \xrightarrow[\text{EtOH}]{\text{organic base}} R\cdot CH{=}CH\cdot CO_2H + CO_2 + H_2O$$

dehydration and decarboxylation follow the condensation. Details are given below for the Doebner modification of this reaction.

Carry out the following reaction with benzaldehyde, *p*-chlorobenzaldehyde, *p*-methoxybenzaldehyde, *p*-hydroxybenzaldehyde, *p*-nitrobenzaldehyde, or cinnamaldehyde. Compare your spectroscopic results with those obtained by other students using different aldehydes.

EXPERIMENT 47. The Doebner reaction

Heat a mixture of the aldehyde (1 g), malonic acid (1.5 g), pyridine (3 ml), and piperidine (2 drops) on the steam bath for one hour, during which time carbon dioxide is evolved. Then boil the mixture for a few minutes, cool, and add to a mixture of crushed ice (40 g) and dilute hydrochloric acid (5 *M*, 20 ml). Filter off the product, wash it with cold water, and recrystallize it from water or ethanol or a mixture of these. Record the yield and m.p. of your product and by comparison with literature values determine whether your product is the *cis* or *trans* isomer.*

Weigh out *accurately* a small amount (20-25 mg) of the acid into a small boat-shaped piece of metal foil. Add the acid and the foil to a little ethanol contained in a graduated flask (100 ml). Shake the flask to dissolve the acid, make up to the mark with ethanol, and shake thoroughly to ensure homogeneity of the solution.

Use a pipette to transfer 1 ml of this solution to a second graduated flask (50 ml) and make the solution up to the mark with ethanol. Shake well to homogenize the solution. Calculate the concentration of this solution in g/litre and in moles/litre.

Record the ultraviolet spectrum of the dilute solution using 1 cm silica cells, and calculate the molar extinction coefficient at the point(s) of maximum absorption. Record the results: λ_{max} ...... nm (ϵ = ...) [see p. 36].

Compare the spectra of cinnamic acid and of 5-phenylpenta-2,4-dienoic acid with that of benzoic acid (Figure 42) and show the effect on λ_{max} of inserting —CH=CH— groups between Ph and CO_2H. Record the results in a table as shown and predict the approximate value of λ_{max} for $Ph(CH{=}CH)_3{\cdot}COOH$.

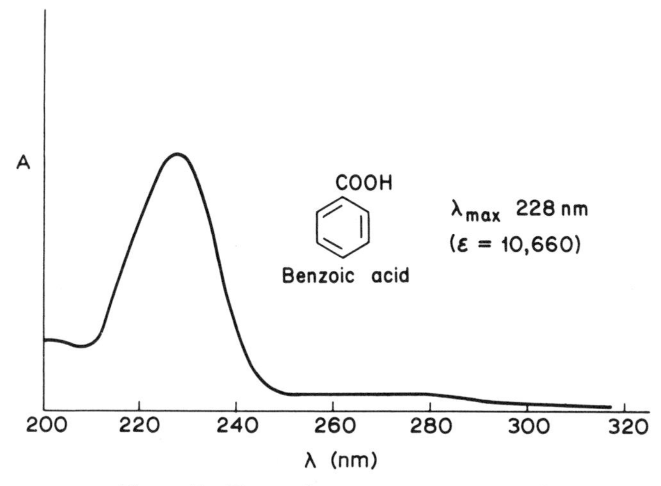

Figure 42. *Ultra-violet spectrum of benzoic acid*

* The stereochemistry of the different isomers of Ph · CH=CH—CH=CH · CO_2H has not been fully established. Two isomers may be obtained from this experiment.

	λ_{max}	ϵ
Ph . COOH		
Ph · CH=CH · COOH		
Ph · $(CH{=}CH)_2$ · COOH		

Compare also the spectra of cinnamic acid and the various substituted cinnamic acids, and comment on the influence of electron-donating substituents (which serve to extend the conjugation) on the position of λ_{max}.

Write an equation for the reaction you have carried out showing the intermediates which are not isolated.

14. A kinetic study

Objective

To illustrate the use of kinetic studies to examine the effect of substituents on the rate of hydrolysis of substituted benzoyl chlorides.

Introduction

[In the following description of this exercise, certain terms used in kinetics are mentioned frequently (e.g. first order, half-life). If these are not understood they should be discussed with the instructor: otherwise it will be difficult to do the experiment correctly.]

The rate of reaction of benzenoid compounds is greatly affected by substituents on the benzene ring. This exercise is designed to illustrate this by determining the rate of hydrolysis of a number of substituted benzoyl chlorides:

$$H_2O + XC_6H_4 \cdot COCl \longrightarrow XC_6H_4 \cdot COOH + HCl$$

The substituent X may be (i) *p*-Br, (ii) *p*-Cl, (iii) *m*-MeO, (iv) *p*-F, (v) H, (vi) *m*-Me, (vii) *p*-Me, or (viii) *p*-Bu^t. Each student should examine one compound and compare his results with those obtained by other students.

Since water is present in great excess, the reaction is pseudo-first order and the rate equation is:

$$\text{rate of reaction} = k[\text{ArCOCl}]$$

Aliquots of the reaction mixture are removed at timed intervals and run into cold acetone. This stops the hydrolysis by cooling and by making the reaction medium less aqueous. The hydrogen chloride liberated is titrated against sodium hydroxide (0.01 *M*). The size of this titre is a measure of the amount of reaction which has occurred. The substituted benzoic acid is so weakly acidic that it does not interfere with the titration. During the titration hydrolysis will still continue although at a much reduced rate, so the titration must be carried out as rapidly as possible. The production of high local concentrations of hydroxide ion is avoided by constant swirling.

EXPERIMENT 48. Hydrolysis of substituted benzoyl chlorides

(i) *Kinetic method.* Before proceeding further, immerse a stoppered flask (100 ml) containing 10% aqueous acetone (50 ml) in a constant temperature bath at 25°. Also immerse a flask containing acetone (*ca.* 200 ml) in ice.

To the aqueous acetone at 25° add the appropriate acid chloride (*ca.* 0.2 g or 0.2 ml, see list given above), shake to dissolve, and start the stop-clock. As the reaction is pseudo-first order, the exact amount added is not important. At timed intervals (see below) remove an aliquot (1 ml), run it into cold acetone (10 ml) contained in a conical flask (50 ml), and rapidly titrate with sodium hydroxide (0.01 *M*) from a burette (10 ml), using lacmoid (B.D.H.) as an indicator. This indicator turns green at the end-point. The length of the timed interval depends on the acid chloride being studied and the following are recommended: *p*-bromo and *p*-chloro, 5 minutes; *m*-methoxy and *p*-fluoro, 9 minutes; benzoyl chloride and its *m*-methyl derivative, 13 minutes; *p*-methyl and *p*-t-butyl, 16 minutes.

Make about eight measurements if possible and enter your results into a table thus:

Time	*Burette reading*		*Titre* (T_t)	($T_\infty - T_t$)	*Log* ($T_\infty - T_t$)
	First	*Second*			

(ii) *Determination of infinity reading* (T_∞). The titre at infinite time, necessary for a first-order kinetic plot, may be determined after the reaction has proceeded for about 10 half-lives. However, this involves a long wait (*ca.* 24 hr in some cases) and an alternative procedure is suggested. In the presence of an excess of hydroxide ions, the hydrolysis of an acid chloride is very rapid. The hydrogen chloride produced neutralizes the alkali so that, to get T_∞, it is necessary to find the amount of sodium hydroxide not neutralized.

Take an aliquot (1 ml) of the reaction mixture and run it into acetone (10 ml). Add an excess of sodium hydroxide (0.01 *M*, 5.0 ml) from a pipette, and allow to stand for 15 minutes before back-titrating the excess of sodium hydroxide with hydrochloric acid (0.01 *M*). Subtraction of this titre from 5.0 gives the value of T_∞.

(iii) *Treatment of results.* Plot a graph of log $(T_\infty - T_t)$ against time. It should be linear and the slope is $k/2.30$. Calculate the value of k. Collect values of k obtained by other students and see which substituents enhance and which diminish the rate of reaction.

L. P. Hammett discovered that, for many reactions like the hydrolysis of benzoyl chlorides, the values of k correlate with the pK of the corresponding acid. Test this with your results by plotting a graph of log k against the pK of the acid (benzoic acid 4.17, *p*-bromo 3.87, *p*-chloro 3.98, *m*-methoxy 4.09, *p*-fluoro 4.14, *m*-methyl 4.27, *p*-methyl 4.37, *p*-t-butyl 4.40).

Why is it unnecessary to determine the initial concentration of acid chloride?

Is the reaction accelerated or retarded by electron-withdrawing groups? Suggest a reason for your conclusion.

15. Nuclear magnetic resonance spectroscopy

Objective

To provide experience in the interpretation of simple n.m.r. spectra, and to use these spectra to help in the identification of unknown organic compounds.

Introduction

The section of Part One on pp. 48-55 sets out the basic principles of n.m.r. spectroscopy, and should be read before the following exercises are attempted. N.m.r. spectra are of use in the identification of organic compounds because they can indicate for any given molecule (a) the number of different environments in which hydrogen atoms occur, (b) the number of hydrogen atoms in each of these environments, and (c) in many cases the nature of that environment, *viz.* the neighbouring atoms or groups of atoms.

EXPERIMENT 49. Simple chemical shifts and spin-spin coupling

Examine the spectrum of **toluene** (Figure 43). Observe the signal at 10 τ which is due to tetramethylsilane (T.M.S.), the reference compound added to the solution to calibrate the chart.

The other two signals are due to resonance of the protons in toluene. Measure the chemical shift of each signal (to the nearest 0.01 τ). Now measure the height of each step in the integrated trace (at the top of the chart). Remember that the integral is proportional to the number of protons in each environment.* Toluene has 5 aromatic protons and 3 methyl protons: which signal corresponds to which set of protons? Check by consulting Figure 39 (p. 51) that you have assigned each signal to the correct protons.

* The accuracy of the integrator may not be better than 10 per cent, but is usually sufficient to permit calculation of ratios of integrals.

Construct a table as shown below and enter the appropriate information for toluene.

TABLE 1

compound	Ar–H (τ)	CH_3-		$-CH_2-$		$-\overset{\mid}{C}H-$	
		τ	*adjacent group*	τ	*adjacent groups*	τ	*adjacent groups*
toluene				—	—	—	—
t-butylbenzene				—	—	—	—
ethylbenzene						—	—
etc.							

Now examine the spectrum of **t-butylbenzene** (Figure 44). From the formula, how many aromatic and how many methyl protons does this molecule possess? Measure the integrals on the chart and decide which signal corresponds to which protons. Enter the τ values in Table 1*.

Does a methyl group attached to a saturated carbon atom give a signal at a higher or lower τ value than a methyl group attached to a π-system?

Now consider **ethylbenzene.** In this molecule there are protons in three different environments: five protons attached to the aromatic ring, two methylene protons, and three methyl protons. Thus the spectrum should consist of three signals. Examine the spectrum of ethylbenzene (Figure 45) and notice that three signals do appear, although only one of these signals consists of a single peak. Measure the integrals of the signals, and use this information to assign each signal to the appropriate protons in the molecules. Enter the chemical shifts† in Table 1.

Examine the $-CH_2-$ and $-CH_3$ signals in detail. How many peaks does each signal contain? Construct Table 2 (p. 129) as shown, and insert the appropriate values for (b) and (c).

Isopropylbenzene also has protons in three different environments: how many protons are in each environment? Examine the spectrum (Figure 46), measure the

* When a signal consists of a complex, unresolved multiplet, the *range* of τ values should be given.

† The chemical shift of protons giving rise to a *symmetrical* multiplet is the mean of the positions of all the component peaks. (If serious distortion of the multiplet occurs, the chemical shift is displaced from the mean position.)

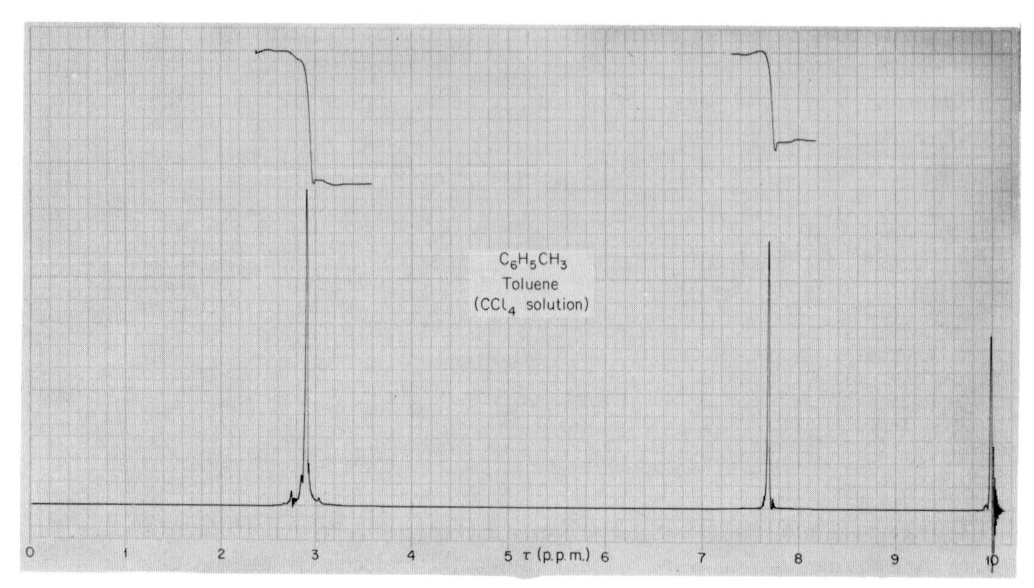

Figure 43. *Toluene*

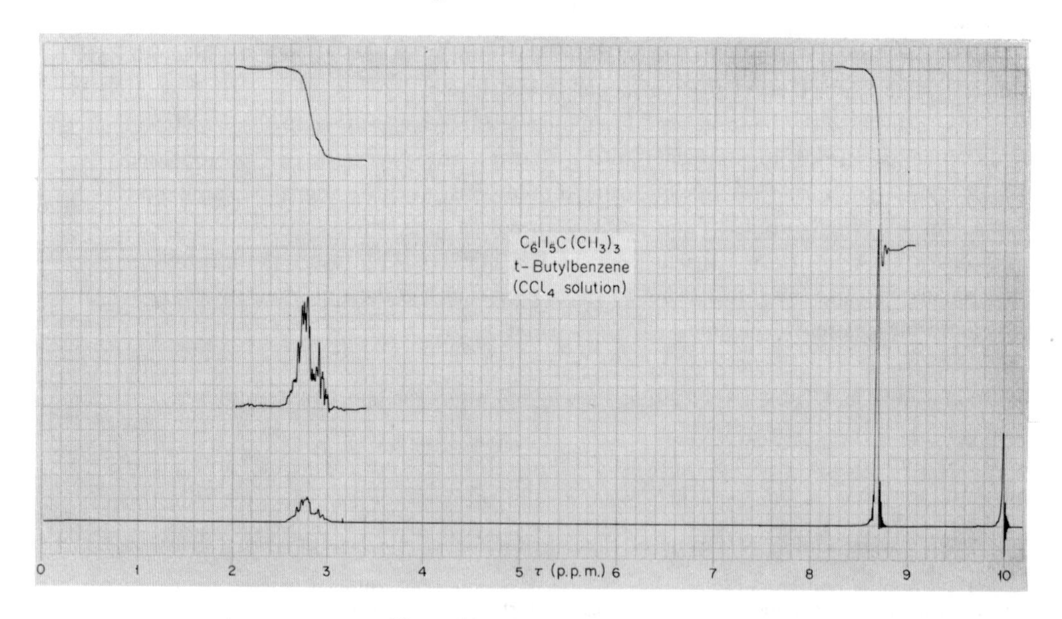

Figure 44. *t-Butylbenzene*

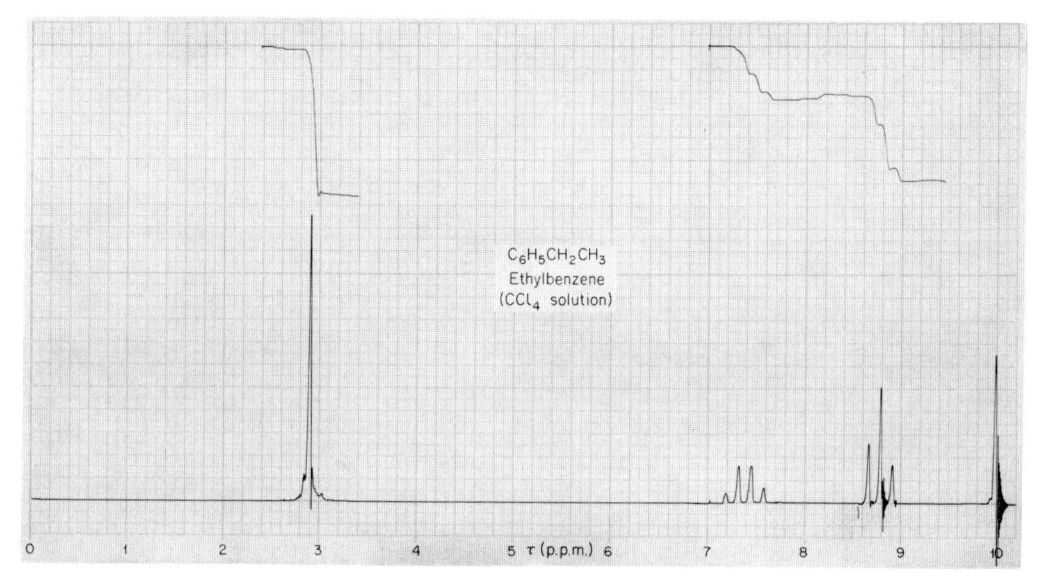

Figure 45. *Ethylbenzene*

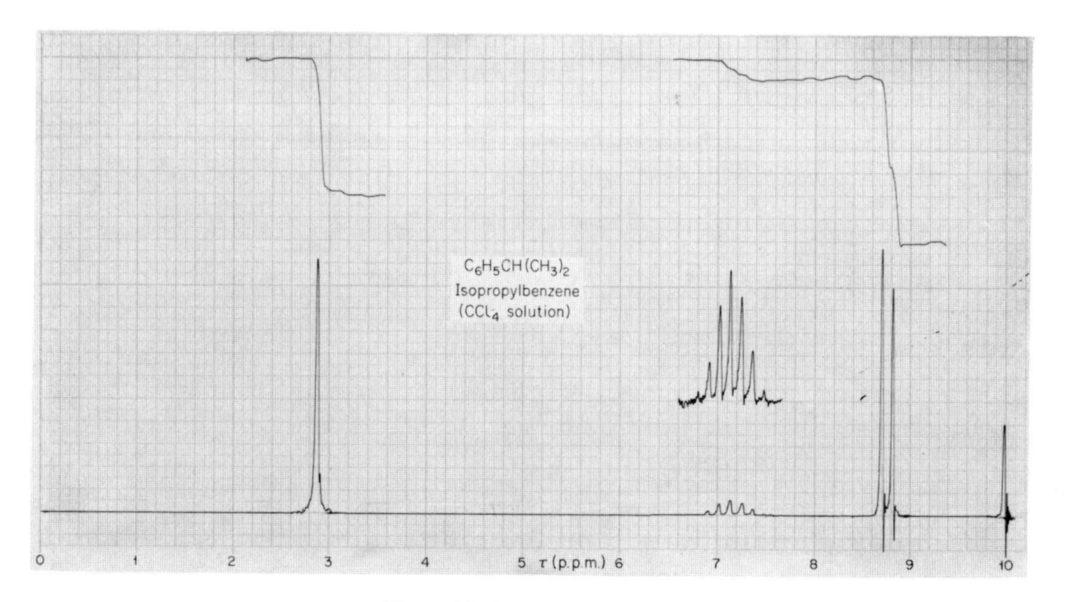

Figure 46. *Isopropylbenzene*

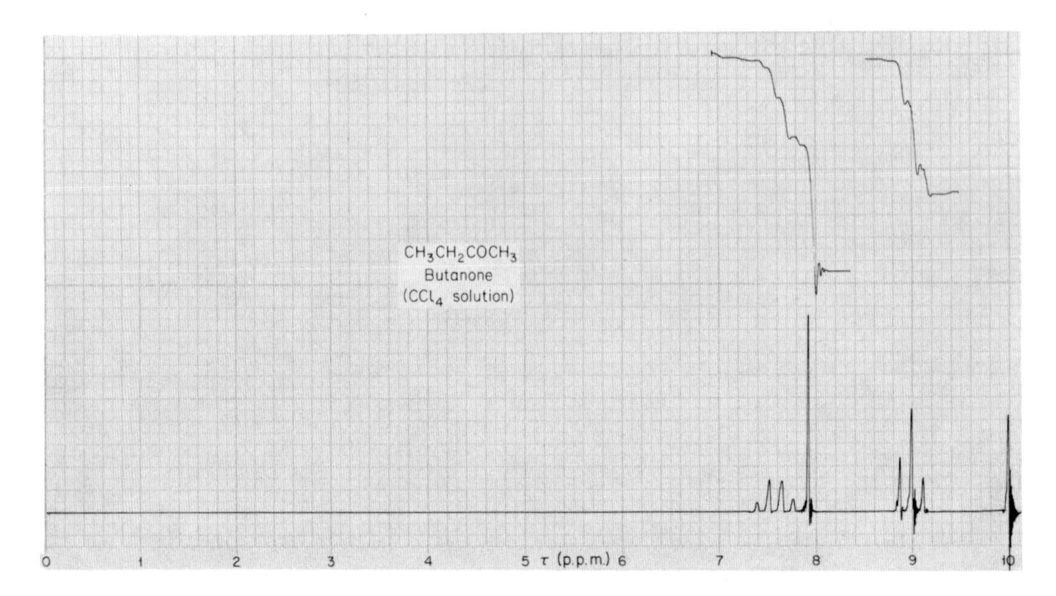

Figure 47. *Butanone*

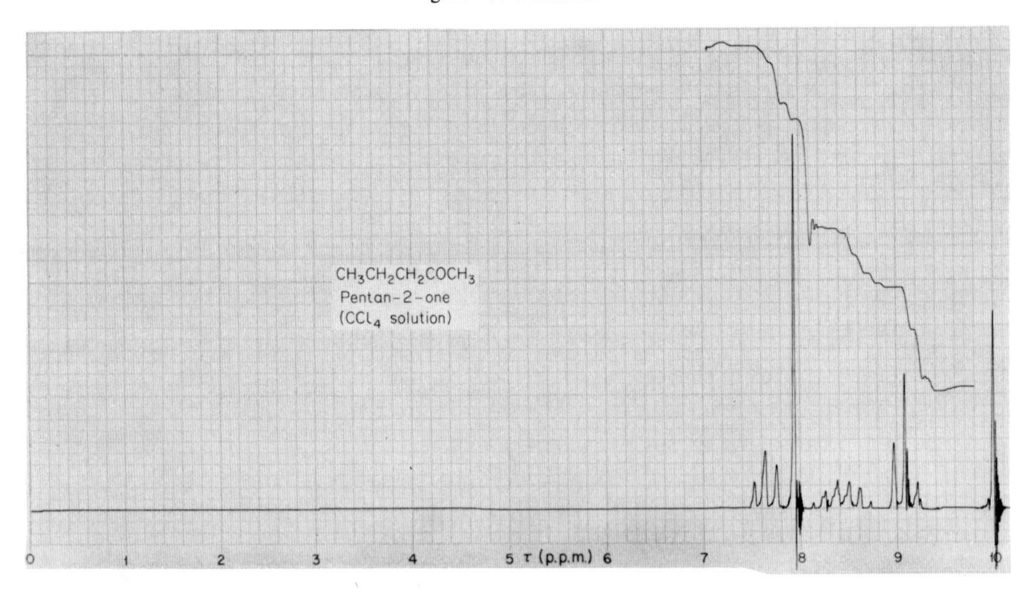

Figure 48. *Pentan-2-one*

TABLE 2

compound	*protons giving signal*	*no. of peaks in signal*	*no. of hydrogens on adjacent carbon(s)*
toluene	Ph–C<u>H</u>$_3$	1	0
		(a)	1
ethylbenzene	PhCH$_2$C<u>H</u>$_3$	(b)	2
ethylbenzene	PhC<u>H</u>$_2$CH$_3$	(c)	3
________	________	(d)	4
________	________	(e)	5*
		(f)	6

* A situation not commonly encountered: but see p. 131.

integrals, and hence assign the signals to the different types of proton. Enter the chemical shifts† in Table 1 and the appropriate data in Table 2, i.e. (a) and (f). (On this spectrum one of the signals is reproduced on a much enlarged scale so that the very tiny outer peaks of the signal are visible.)

If your entries in Table 2 are correct (check them with your instructor) you should be able to see a relationship between the number of peaks in the signal and the number of protons on the adjacent carbon atom(s), and be able to fill in the two missing numbers, (d) and (e), in the third column of Table 2.

Can you now predict the appearance of the spectrum of 1,3-diphenylpropane ($PhCH_2CH_2CH_2Ph$)? Consider the following points:

(a) In how many different environments do protons occur?

(b) How many protons occur in each environment?

(c) What is the approximate chemical shift of each signal? Use Table 1, and Figure 39 (p. 51) to help with this.

(d) How many peaks will appear in each signal? Count the number of protons on adjacent carbon atoms, and use Table 2.

† See footnote p. 125 †

Try to draw this spectrum in your notebook; discuss it with your instructor.

Consider the structure of **butanone**. It contains protons in three different environments. Which of these will give rise to a single peak in the n.m.r. spectrum, and which will be split into groups of peaks? Look at the spectrum (Figure 47) and identify the signal from each set of protons. Enter the results in Table 1.

Repeat the exercise with **ethyl acetate** (Figure 40, p. 52) and **methyl acetate** (Figure 38, p. 49), and note the differences in chemical shift between alkyl and alkoxy groups.

Can you suggest how the spectrum of methyl propionate ($CH_3CH_2COOCH_3$) will appear?

EXPERIMENT 50. Coupling constants and spectra of aromatic compounds

In the previous experiment, no attention was paid to the spacing between member peaks of a multiplet. This spacing is called the **coupling constant**, *J*, and is usually expressed in Hz (on this chart 1 small division represents 6 Hz).

The coupling constants for interactions between adjacent protons in aliphatic compounds (where free rotation can occur about single bonds) are all approximately equal. Measure the spacing between the peaks of typical multiplets (e.g. the quartets and triplets in Figures 45 and 47) and express the coupling constant in Hz.

Now examine the spectrum of **pentan-2-one** (Figure 48) and try to identify the various signals using Table 1 if necessary. Tabulate the results as shown (Table 3).

TABLE 3

protons	*chemical shift (τ)*	*multiplicity*	*integral*
$C\underline{H}_3CH_2CH_2COCH_3$		Triplet	3H (——chart divisions)
$CH_3C\underline{H}_2CH_2COCH_3$			2 H
$CH_3CH_2C\underline{H}_2COCH_3$			2 H
$CH_3CH_2CH_2COC\underline{H}_3$			3 H

You have probably written 'sextet' for the multiplicity of the $CH_3C\underline{H}_2CH_2COCH_3$ signal. This implies (cf. Table 2) that these two protons are adjacent to five equivalent protons. This, however, is clearly not so: the $C\underline{H}_3CH_2$ protons and the $CH_2C\underline{H}_2CO$ protons are not in the same environment. The $CH_3C\underline{H}_2CH_2$ signal should be split into a quartet by the methyl protons on the one side and into a triplet by the methylene protons on the other side, and should thus appear as twelve peaks (either three quartets or four triplets), as shown in Figures 49(a) and 49(b).

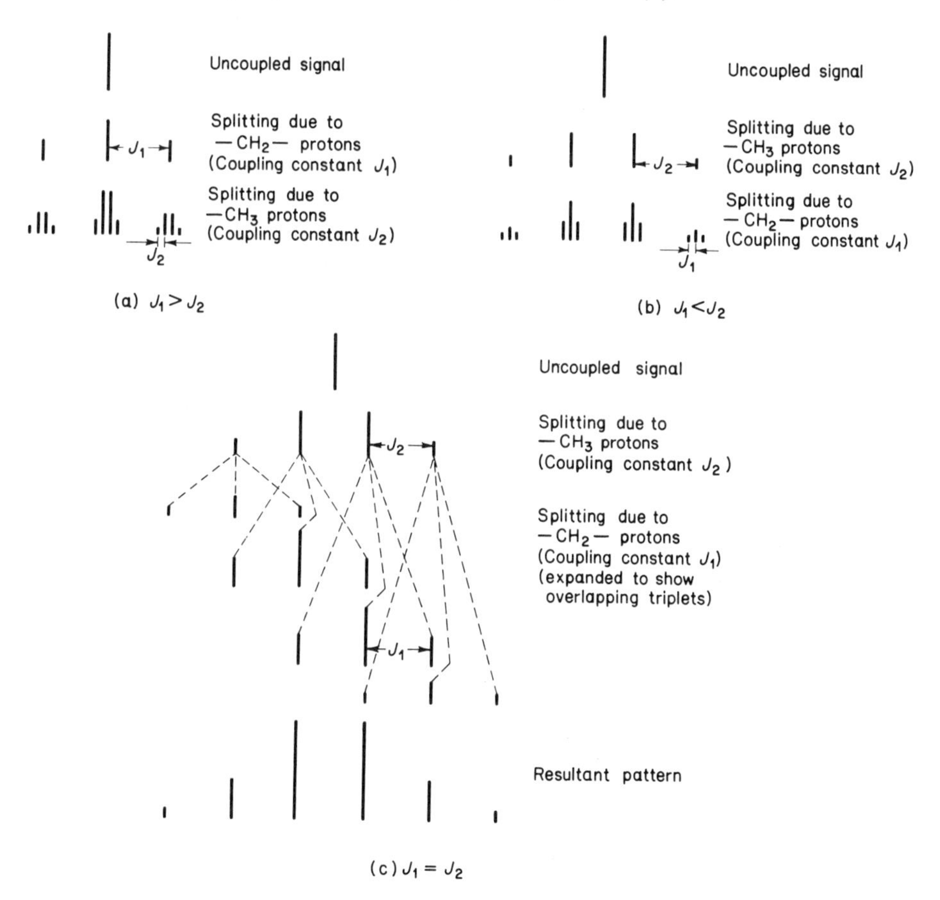

Figure 49.

Pentan-2-one is a case where J_1 is approximately equal to J_2 and so the signal is approximately a sextet as in Figure 49(c).

Now examine the spectrum of **vinyl acetate** (Figure 50). Identify the signal due to the methyl group, and notice that the three protons of the vinyl group give rise to signals at considerably lower τ values than protons attached to saturated carbons. Also the simple relationship between the multiplicity of signals and the number of protons on the adjacent carbons, which was established in Table 2, is no longer valid: the $\mathrm{C\underline{H}_2{=}C}\!<_{\mathrm{H}}$ signal is not a doublet, but consists of four doublets, and the $\mathrm{CH_2{=}\acute{C}\underline{H}}$ signal is not a triplet (1:2:1) but four peaks of equal intensity.

The additional peaks in the spectrum result from the non-equivalence of the two $CH_2{=}$ protons. Because free rotation about the double bond is impossible, proton H_a is fixed in the position *trans* to H_c, and H_b is fixed in the position *cis* to H_c; thus H_a and H_b are in different environments.

$$\begin{array}{ccc} CH_3COO & & H_a \\ & C{=}C & \\ H_c & & H_b \end{array}$$

Coupling constants for *trans* protons are large (13-18 Hz: 2-3 chart divisions) and those for *cis* protons are slightly smaller (6-12 Hz: 1-2 chart divisions). The non-equivalent protons on the same carbon atom (e.g. H_a and H_b) give rise to very small coupling constants (0.5-3 Hz: $<\frac{1}{2}$ chart division).

Work out the expected splitting patterns for the H_a, H_b, and H_c signals. The H_a signal will be split into a widely spaced doublet by the *trans* H_c proton, and each element of this doublet will be further split into a doublet (very closely spaced) by H_b. So the H_a signal should look like this:

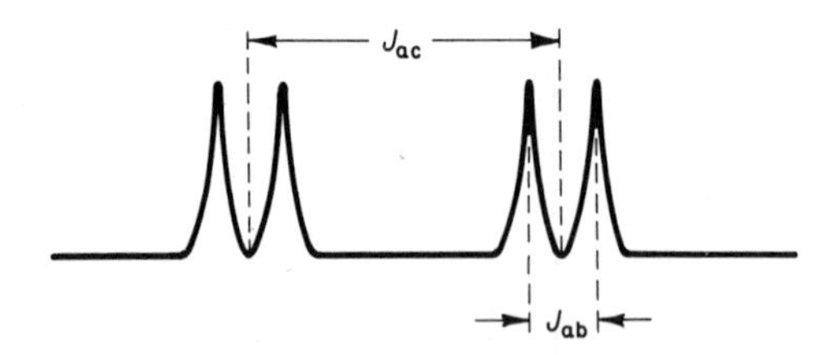

Work out the expected splitting patterns for the H_b and H_c signals in the same way. Look again at the spectrum. Can you identify the signals from H_a, H_b, and H_c? Tabulate the τ values for each signal, and measure J_{ab}, J_{bc}, and J_{ac}.

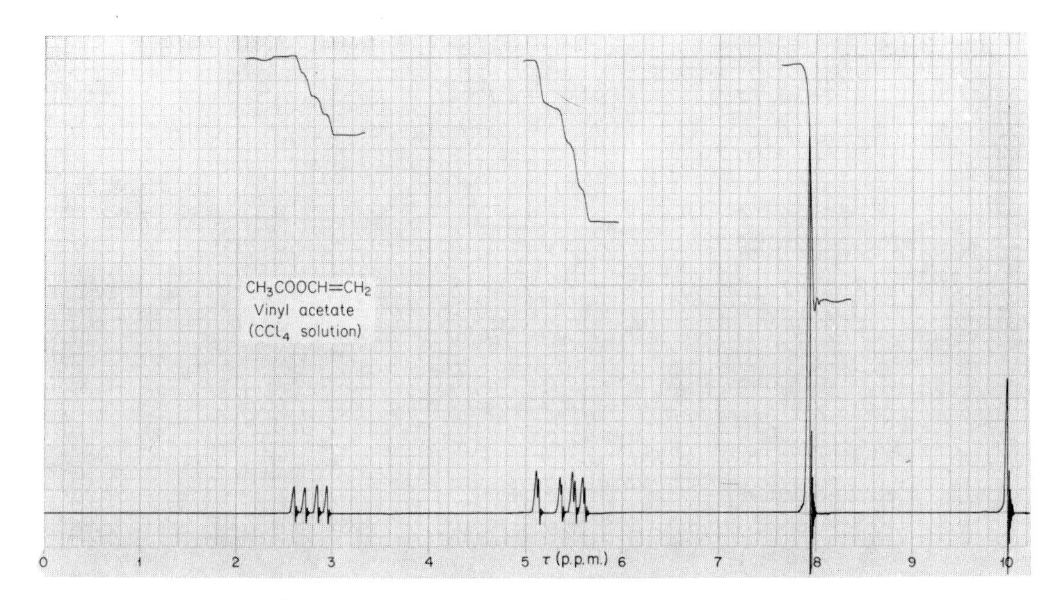

Figure 50. *Vinyl acetate*

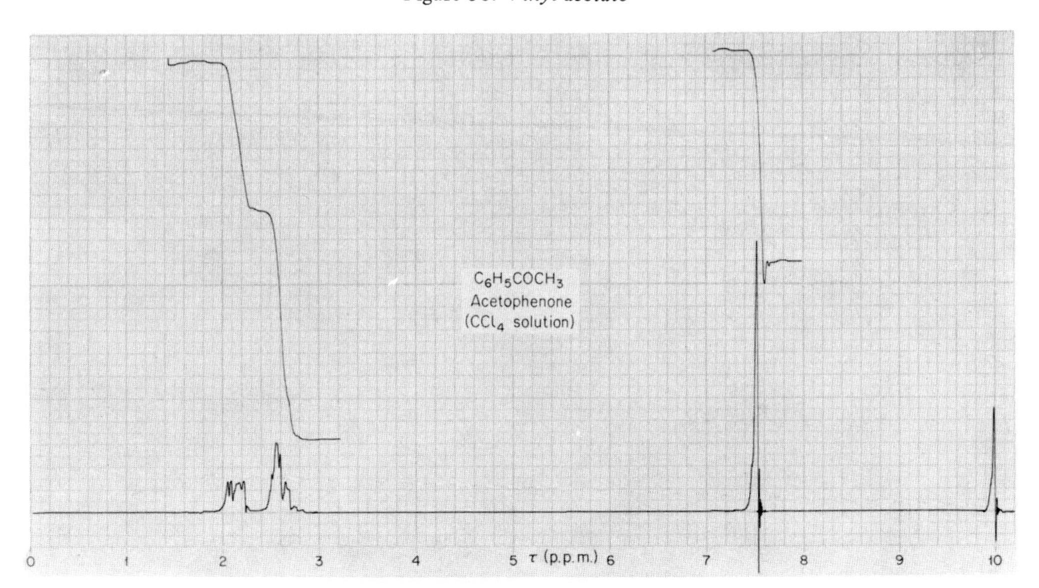

Figure 51. *Acetophenone*

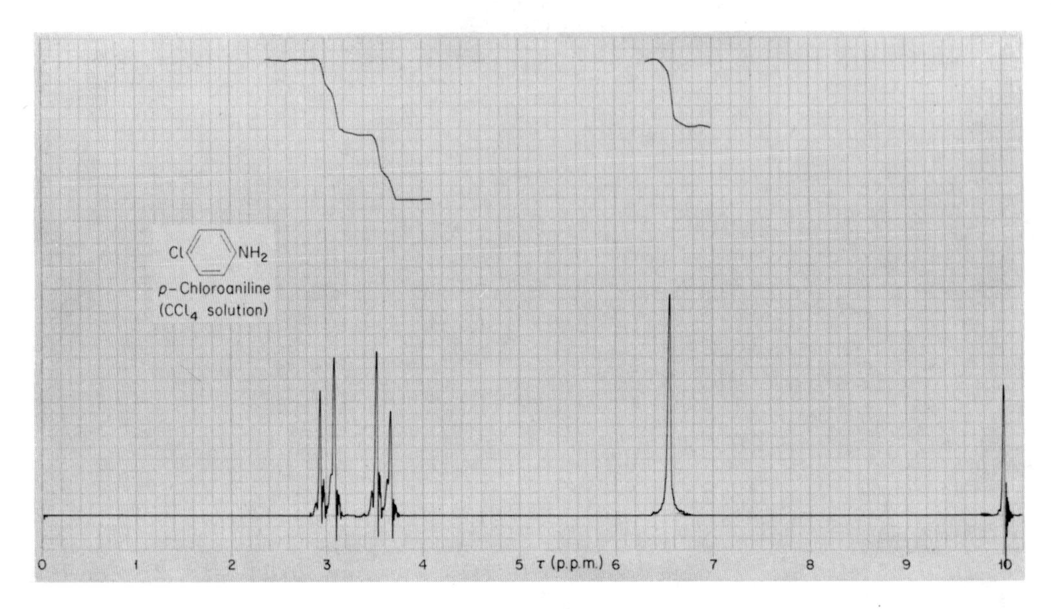

Figure 52. p-*Chloroaniline*

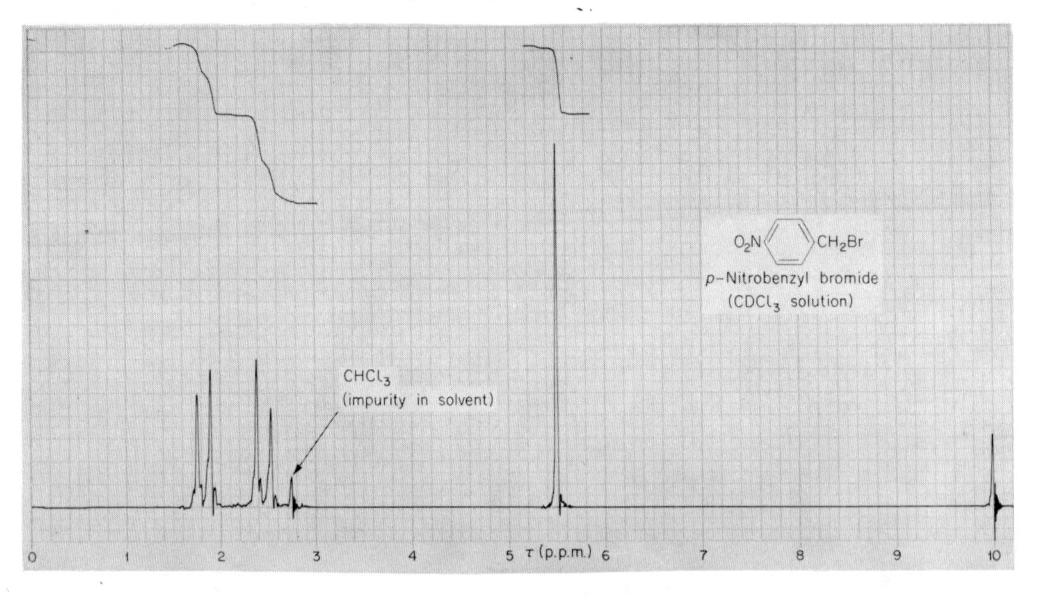

Figure 53. p-*Nitrobenzyl bromide*

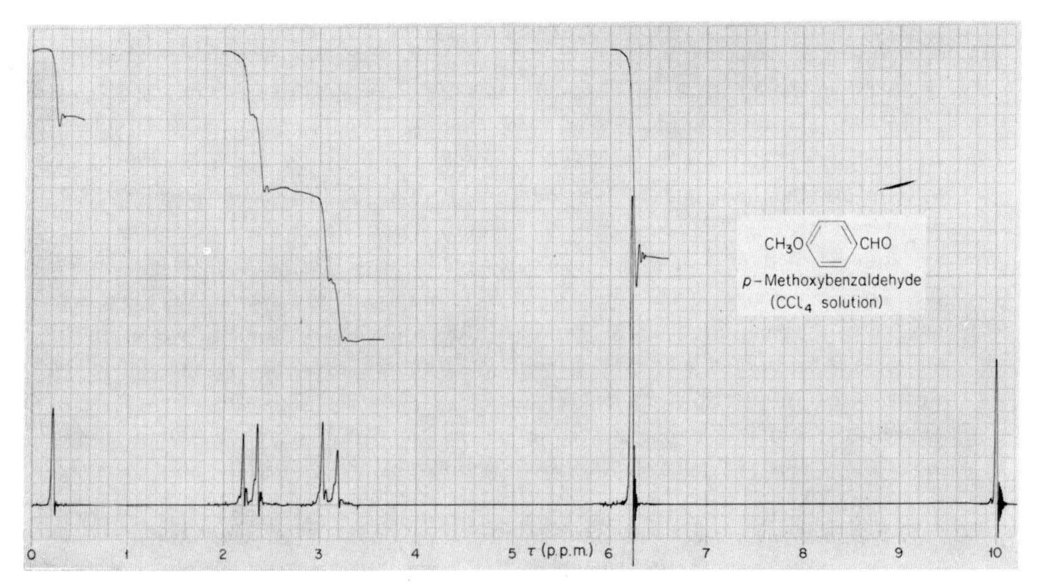

Figure 54. p-*Methoxybenzaldehyde*

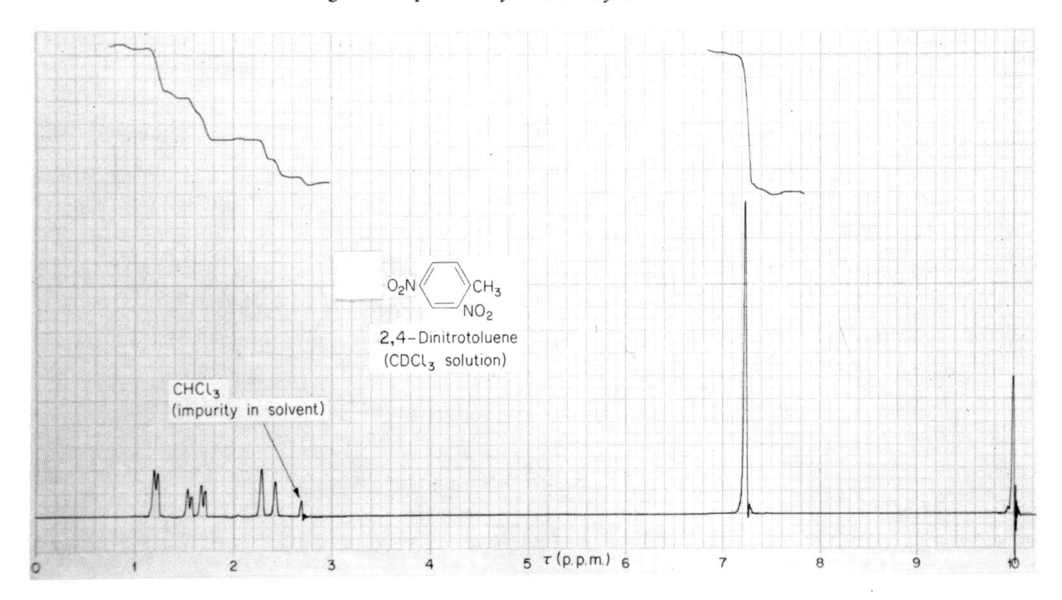

Figure 55. *2,4-Dinitrotoluene*

So far in these experiments little account has been taken of the n.m.r. spectra of protons attached to an aromatic ring, except to note the approximate chemical shift of Ar—H signals (Table 1). It is unusual for Ar—H signals to give rise to singlets (as do toluene, ethylbenzene, and isopropylbenzene, Figures 43, 45, 46); complex multiplets (e.g. t-butylbenzene and acetophenone, Figures 44 and 51) are much more commonly encountered. Although such multiplets are usually extremely difficult to resolve into their component signals, they often yield useful information. For example, the overall integral of the multiplet is an indication of the proportion of aromatic protons in the molecule. Also the range of τ values over which the multiplet occurs is related to the electron-rich or electron-deficient character of the ring. Compare the spectra of **t-butylbenzene** and **acetophenone**. Does an electron-withdrawing substituent raise or lower the τ value of the signal? Electron-releasing substituents have the opposite effect (cf. Figure 39, p. 51).

Sometimes (e.g. in acetophenone) a multiplet may be split into two distinct groups. Measure the integrals for the two groups of aromatic protons in the acetophenone spectrum. How many protons are in each group? Can you suggest why the x protons have a significantly different environment from the remaining $(5 - x)$?

Three special cases are worthy of further mention, since the aromatic signal may be more fully analysed.

(1) The appearance of the aromatic signal as a *singlet* indicates that all the aromatic protons are in the same or very similar environments. Such absorption is characteristic of many monosubstituted aromatic compounds where the substituent is not strongly electron-releasing or withdrawing (Figures 43, 45, 46) and also of symmetrically *para*-disubstituted compounds.

(2) A *symmetrical multiplet* is indicative of certain elements of symmetry in the substitution pattern of the aromatic ring. In particular unsymmetrically *para*-disubstituted compounds (e.g. ***p*-chloroaniline**, Figure 52) usually give a symmetrical group consisting of four main peaks together with a number of subsidiary peaks.*

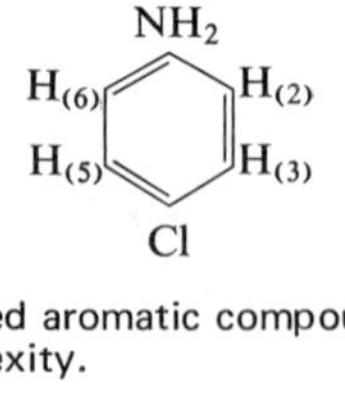

NH_2

$H_{(6)}$ $H_{(2)}$

$H_{(5)}$ $H_{(3)}$

Cl

* Symmetrically *ortho*-disubstituted aromatic compounds also give rise to a symmetrical Ar—H multiplet, but of much greater complexity.

Use Figure 39 (p. 51) and your conclusions above about the acetophenone spectrum to predict which half of the *p*-chloroaniline multiplet corresponds to the signal from $H_{(2)}$ and $H_{(6)}$, and which half corresponds to $H_{(3)}$ and $H_{(5)}$. Repeat this exercise with the spectra of ***p*-nitrobenzyl bromide** (Figure 53) and ***p*-methoxybenzaldehyde** (Figure 54). Arrange the substituents in order according to the chemical shift of the adjacent protons (in ascending or descending order of τ values, as you wish). Does this order bear any relationship to the order of electron-withdrawing (or releasing) power of the substituents?

Note the chemical shifts of amino-protons (Figure 52) and aldehydic protons (Figure 54).

(3) In a few instances, all of the aromatic protons are in sufficiently different environments for complete analysis of the signal to be made. An example of this is **2,4-dinitrotoluene** (Figure 55). Work out from the formula which of the aromatic protons is in the most electron-deficient environment, and which is in the least electron-deficient environment. Which signal should have the lowest τ value, and which the highest?

CH_3
$H_{(6)}$ NO_2
$H_{(5)}$ $H_{(3)}$
NO_2

The coupling constants for interaction between Ar–H nuclei are given in the table on p. 55. Coupling between *para*-protons gives rise to such small J values that splitting is not observed except under very high resolution.

Now work out the theoretical appearance of each signal (as you did earlier for vinyl acetate). $H_{(3)}$ is *meta* to $H_{(5)}$ and so the $H_{(3)}$ signal appears as a closely spaced doublet (further splitting due to the *para*-situated $H_{(6)}$ not being observable).

Consider the $H_{(5)}$ and $H_{(6)}$ signals in a similar way. Can you identify these signals on the spectrum? Measure $J_{(5)\text{-}(6)}$ (i.e. J_{ortho}) and $J_{(3)\text{-}(5)}$ (i.e. J_{meta}).

EXPERIMENT 51. N.m.r. spectra of unknown compounds

The following examples provide practice in the interpretation of n.m.r. spectra. Use Figure 39 and the reference compounds on the preceding pages to help, if necessary, in the identification of **X**, **Y**, and **Z**.

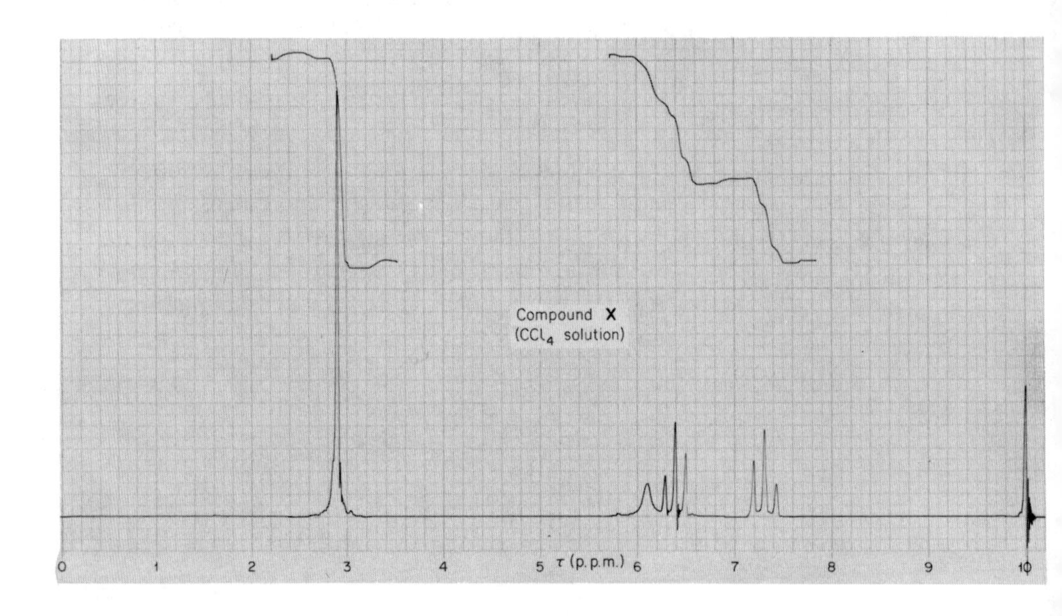

Figure 56. *Compound* X

Compound **X** shows strong O—H absorption in the infrared spectrum.

Compound **Y** shows strong C=O absorption in the infrared spectrum ($\bar{\nu}_{max}$ 1720 cm^{-1}). Suggest *two* possible structures for **Y**.

Compound **Z** shows strong O—H absorption in the infrared spectrum. The singlet at 6.45τ overlaps with a multiplet; the multiplets at 8.58 and 9.09τ are considerably distorted (cf. Figure 48).

When you have completed this exercise, your instructor may issue an n.m.r. spectrum along with each new compound for identification.

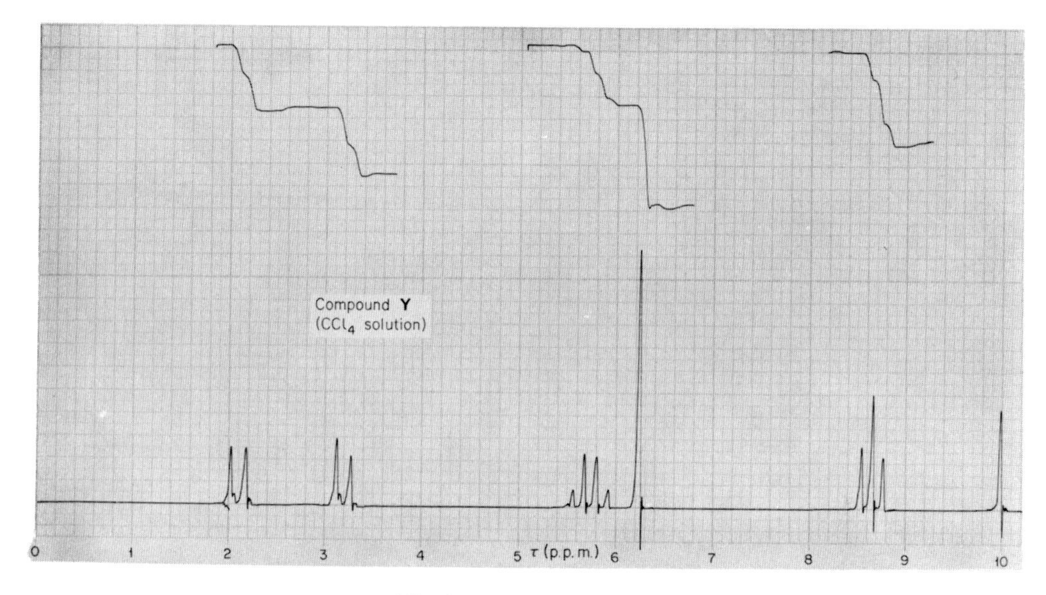

Figure 57. *Compound* Y

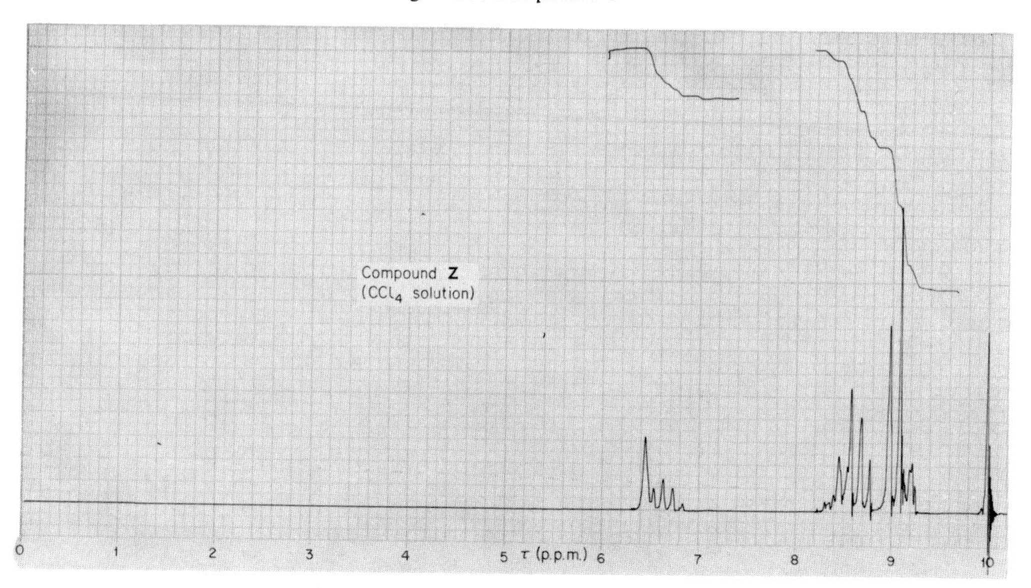

Figure 58. *Compound* Z

PART THREE

The identification of organic compounds

Introduction

We are concerned here with the identification of organic compounds which have been reported before. A trained and experienced chemist obtains clues in various ways, but final proof rests on an identity of properties between the unknown compound and either an authentic sample available for comparison or one whose properties are already known. Melting point has been the most-used point of identity, but identity of spectroscopic and chromatographic properties are becoming increasingly acceptable.

In a research laboratory this identity is established with the minimum of effort and time consistent with the maximum of certainty. Short cuts are permitted so long as the final proof is satisfactory, and mixed melting points are not considered as cheating. Spectroscopic information reduces the necessity for carrying out many of the chemical tests formerly studied systematically.

We believe that these methods should be permitted and encouraged in undergraduate laboratories. If a student thinks he recognizes his unknown compound from its smell or appearance, and is able to prove this by melting point or mixed melting point of the compound or a derivative and/or by spectroscopic and chromatographic evidence, then the exercise is completed and the next one can be undertaken. We accept that, occasionally, the student should try additional tests to get experience of these, but the emphasis throughout is on finding and interpreting clues and on comparing unknown with known compounds. Tests should be selected carefully and interpreted critically. In this way the student learns to pit his wits and his skills against the chemical nature of his compounds and not against the instructor. The instructor's aim is to teach his students how to perform the tests, how to assess them, and how to build them into proof which is both elegant and complete.

The methods used to identify organic substances do not have the sequential logic of those used in the study of metallic ions. Identification requires careful experimentation and informed interpretation: thought as well as action.

The investigation generally falls into two parts: (i) recognition of the functional groups present and (ii) identification of the compound.

Functional groups are recognized by a study of spectroscopic, chromatographic, and other physical properties along with appropriate chemical tests.

Identification involves a comparison of the properties of the unknown with those of a known compound. When the latter is available or easily made, direct comparison is possible; otherwise a comparison is made between the observed properties of the unknown and the reported properties of the known compound. The properties most commonly compared are melting points and infrared spectra; identity of chromatographic properties may also be used. *It is essential to establish more than one point of identity before coming to a final conclusion.*

Practical points

The results obtained for each test must be carefully recorded immediately the test is completed.

These tests can be satisfactorily carried out on the scale indicated so long as reagents of appropriate concentration are used in the recommended amounts.

Try the tests with known compounds whenever there is any doubt about the nature of the expected results. Your instructor will help you to select suitable compounds.

Follow the instructions in sections 1-6 then select carefully among the remainder those which seem to be of value. It is easier to do additional simpler tests than to prepare derivatives, so it is worth checking your conclusions before making derivatives.

1. Purification

Since some tests are easily vitiated by small amounts of impurities the unknown compound should first be purified. This is particularly necessary when samples are taken from bottles which have been in use for some time. Amines and phenols oxidize quickly and carbonyl compounds undergo oxidation or condensation.* Solids should be recrystallized (p. 3) and liquids distilled (p. 9), under reduced pressure if necessary. Whilst doing this the melting point of a solid and the boiling point of a liquid should be measured and recorded.

* Instructors should not distribute samples of alkenes or ethers which may be peroxidized as these can be purified only with special precautions.

2. Appearance and smell

Record these but unless very experienced take care not to be misled.

Substances which are still strongly coloured after purification probably contain one or more of the chromophoric groups NO, NO_2, or N=N, and possibly an additional salt-forming group such as a phenol, amine, or sulphonic acid. Colour is enhanced by increasing delocalization of electrons, so that nitro-phenols are more coloured in alkaline solution and nitro-amines are less coloured in acidic solutions. (Discuss this with your instructor if you do not understand it.)

3. Spectra

Obtain the infrared spectrum of the compound to be identified and examine it carefully (see p. 25). Pay particular attention to the possible occurrence of absorption bands due to the following groups: $>C=O$, O–H, N–H, and NO_2. Wherever possible, functional groups detected spectroscopically should be confirmed by the appropriate chemical tests.

Also examine the n.m.r. spectrum if available. This may be more informative when the functional groups present have been recognized (see pp. 48 and 124).

4. Detection of nitrogen, sulphur, and halogens

The presence of nitrogen, sulphur, and halogens can be detected by Lassaigne's sodium fusion test. The major compounds containing these elements are:

nitrogen: amines, amides, nitriles, nitro compounds.
sulphur: sulphonic acids and derivatives, sulphoxides and sulphones, thio compounds.
halogens: acyl, alkyl, and aryl halides.

The more common compounds which do not contain these elements are hydrocarbons, alcohols, ethers, and all carbonyl compounds. Remember that an unknown compound may have two or more functional groups.

Test. Fusion. Place a small piece of sodium (*ca.* 3 mm cube) in a small test tube, supported by a clamp, a pair of tongs, or a wooden clothes peg. Just melt the sodium by means of a small Bunsen flame, then add the unknown substance (20 mg) so that it drops directly on to the molten sodium.* (Take care, especially with liquids, not to add too much material.) Heat the tube, gently at first then more strongly, until the end of the tube is red hot. Whilst still hot, plunge it into water (2 ml) contained in a test tube. Filter the hot solution and boil the residue of charred material and broken glass with more water (2 ml). The filtrate, which should be water-clear and alkaline, may contain sodium cyanide, sulphide, and halides. It may be necessary to carry out five tests with this solution.

(i) *Sulphur.* Dilute one drop of the filtrate and add freshly-prepared sodium nitroprusside solution; a purple coloration indicates the presence of sulphur.

(ii) *Nitrogen.* Add some of the filtrate (1 ml) to a tube containing powdered ferrous sulphate (100 mg), heat gently until the solution boils, and acidify the hot solution with dilute sulphuric acid. A blue precipitate or coloration of ferric ferrocyanide (Prussian blue) proves that nitrogen is present. (If sulphur is present, a black precipitate results on addition of ferrous sulphate, and hydrogen sulphide is evolved when the solution is acidified.)

(iii) *Halogens.* Add a small excess of silver nitrate solution to the alkaline solution (1 ml) when a brown or black precipitate will be formed. Add excess of concentrated nitric acid and warm the mixture. Hydrogen cyanide and hydrogen sulphide are liberated if nitrogen and sulphur were originally present. Boil the mixture: if no precipitate survives halogen is absent. Decant the supernatant liquid and re-boil any precipitate with fresh nitric acid. Any precipitate remaining is silver halide. Decant the acid solution, wash the precipitate with water, and test its solubility in concentrated ammonia solution. Silver chloride is white and easily soluble in ammonia, the bromide is pale yellow and soluble only with difficulty, the iodide is yellow and insoluble.

(iv) If the above test is not satisfactory or if more than one halogen is present, proceed in the following way. Add chlorine water and carbon tetrachloride to the alkaline solution (0.8 ml) and observe whether any colour develops in the lower layer. This will be purple if iodine is present and brown if bromine is present (often not easily detected) whilst chlorine gives no coloration

(v) Chlorine and bromine are distinguished by heating the alkaline filtrate (0.8 ml) with a little potassium permanganate solution and dilute sulphuric acid whilst

* Alternatively, the unknown substance and the sodium may be placed together in the tube before heating.

holding a piece of moist fluorescein-paper over the mouth of the tube. If the paper turns pink, bromine is present; chlorine bleaches the paper or causes little visible change; iodine may darken the paper but is better detected by the first test.

5. Action of heat

Test. Heat a little of the substance on a metal spatula or crucible lid. Note whether it burns, whether the flame is smoky, clear yellow, or blue, and whether there is any residue after prolonged heating.

Comment. Aromatic compounds generally burn with a smoky flame and aliphatic compounds with a blue or clear yellow flame. The residue may be metallic, arising from the salt of an acid or phenol or from the bisulphite compound of an aldehyde or ketone.

6. Detection of acidic and basic groups

It is possible to determine whether a compound contains acidic and/or basic groups by simple solubility tests. These are based on the fact that (i) most organic compounds are insoluble in water, (ii) basic organic compounds are more soluble (and usually dissolve) in aqueous acidic solutions, and (iii) acidic organic compounds are more soluble (and usually dissolve) in aqueous alkaline solutions. The test is not appropriate for water-soluble compounds because changes in solubility are then not easily detected. Aqueous solutions of such substances can, however, be tested with litmus or other indicators. Amphoteric substances may show increased solubility in acid and in alkali.

The most important acidic compounds are carboxylic and sulphonic acids, phenols, and sulphonamides. Basic substances are most commonly amines.

Test. Place one drop of liquid or a few crystals of powdered solid in each of three test tubes (75 x 10 mm) and add, respectively, 1 ml of water, dilute hydrochloric acid (2 *M*), and dilute sodium hydroxide (2 *M*). Shake the tubes and observe whether the substance dissolves in the cold.

If the substance dissolves in water, test the solution with litmus paper. Also test its solubility in dry ether (Test 8) but not in dilute acid or alkali.

Comment. Some amines are so weakly basic that they dissolve only in concentrated hydrochloric acid. This reagent, however, may also dissolve some alcohols and carbonyl compounds which are very weak bases. Some amines form insoluble salts which are precipitated on addition of acid.

If the compound appears to be basic, it must almost certainly be an amine and should therefore contain nitrogen. If these two tests do not conform, check both results. Amines should be tested with nitrous acid (Test 12).

The preliminary examination is now complete. Among the following tests (7-18) only those which seem appropriate need be undertaken. The selection must be made thoughtfully and where there is any doubt the instructor should be consulted.

7. Further tests for acidic compounds

Reactions with sodium bicarbonate solution and 'neutral' ferric chloride solution allow a distinction to be made between carboxylic acids and phenols.

(i) *Test.* Add the unknown substance (20 mg) to a saturated aqueous sodium bicarbonate solution and observe whether carbon dioxide is liberated. If the unknown is solid and appears to be insoluble, add one or two drops of ethanol.

Comment. Carboxylic acids, acid chlorides and anhydrides, readily-hydrolysed esters, sulphonic acids and sulphonyl chlorides, and nitrophenols give a positive reaction. Other phenols do not. Salts of weak bases may give a positive test. Insoluble acidic compounds may be very slow to react.

(ii) *Test.* Neutralize ferric chloride solution by adding dilute sodium hydroxide solution dropwise until a small permanent precipitate of ferric hydroxide is just formed. Separate and use the clear liquor. If too much alkali is added, no ferric ion may remain.

Shake the unknown (5 mg) with water (0.5 ml), add one drop of the neutralized ferric chloride solution, and observe any change in colour. Repeat the test using an ethanolic solution of the unknown.

Comment. The appearance of a colour, which may be transient, indicates the presence of a phenol or enol (e.g. a β-diketone). α-Hydroxy-acids give a strong yellow colour and carboxylic acids sometimes give coloured precipitates.

8. Water-soluble compounds

Substances soluble in cold water and in ether are usually polar compounds of low molecular weight. These include the lower alcohols, aldehydes, ketones, acids, esters, amides, nitriles, and some phenols. The infrared spectrum should be informative and the compounds should give the expected chemical tests.

Substances which are soluble in cold water but insoluble in ether are highly polar compounds, including salts of carboxylic acids, of sulphonic acids, and of nitrogen bases, along with carbohydrates, hydroxy acids, polybasic acids, sulphonic acids, some aminosulphonic acids, some amides, some ureas, thiourea, and chloral hydrate. Some restriction of this list is possible depending on whether the compound is neutral, basic, or acidic.

9. Carbonyl compounds (aldehydes, ketones, and carboxylic acids and their derivatives)

The C=O bond is easily recognized from the infrared spectrum, and compounds containing this group should be tested with 2,4-dinitrophenylhydrazine. A positive test shows the presence of aldehyde and/or ketone (the carboxyl group and its derivatives may also be present). A negative test shows the absence of aldehydes and ketones. Some tests for distinguishing between aldehydes and ketones are given: these are not always satisfactory and seldom necessary since the characterizing derivatives of aldehydes and ketones are the same.

The presence of a carboxyl group will be apparent from the infrared spectrum (p. 30) and the acidity of the compound (Tests 6 and 7). Acyl halides and amides may be present if the infrared spectrum shows carbonyl absorption and if the sodium fusion test indicates the presence of the appropriate element. Esters, amides, and nitriles may have to be hydrolysed and the products (carboxylic acid along with alcohol, phenol, amine, or ammonia) identified (see section 34).

9a. Test for aldehydes and ketones

To the unknown (1 drop of liquid or 20 mg of solid dissolved in the minimum volume of methanol) add Brady's reagent (1 ml, 2,4-dinitrophenylhydrazine sulphate in methanol).* If no precipitate appears on shaking, boil for one minute and cool. A yellow, orange, or red precipitate is positive.

* See footnote, p. 74.

Comment. The colour of the precipitate depends on the extent of conjugation in the product. Aliphatic derivatives are usually yellow or orange; aromatic components and derivatives of $\alpha\beta$-unsaturated carbonyls generally yield orange or red precipitates. Acetals, oximes, and other compounds hydrolysed to aldehydes or ketones may also form 2,4-dinitrophenylhydrazones.

Under the conditions of this reaction amines are often precipitated as their sulphates which, though normally white, may appear yellow in this solution. Do not confuse this with a precipitate of a 2,4-dinitrophenylhydrazone. If in doubt, filter the precipitate and test its solubility in water: amine salts are generally soluble.

9b. Supplementary tests for aldehydes and ketones

(i) *Test.* Dissolve the compound (10 mg) in a few drops of ether, shake with a little saturated sodium bisulphite solution, and notice whether any crystalline product separates.

Comment. Aldehydes, most alkyl methyl ketones (but not aryl methyl ketones), cycloalkanones, and α-diketones readily yield crystalline bisulphite compounds.

(ii) *Test.* Place silver nitrate solution (1 ml, 0.3 *M*) in a thoroughly clean test tube, add a drop of dilute sodium hydroxide (1 *M*) and then dilute ammonia (1 *M*) until the precipitate just redissolves. Do not use excess of ammonia solution. Add the unknown substance (10 mg) to this solution and shake. If there is no reaction in the cold, warm gently, but do not boil. A silver mirror or precipitate of silver is positive.

Comment. Aldehydes produce a silver mirror; ketones (except α-diketones and α-hydroxyketones) do not. Some other reducing substances of quite different nature such as hydroxylamines and salts and esters of formic acid also reduce silver oxide to silver. Insoluble compounds may react only slowly.

(iii) *Test.* Add the unknown (10 mg) to Fehling's solution (0.5 ml of solutions A and B), warm, and see if there is any red-brown precipitate of cuprous oxide.

Comment. Aldehydes and α-hydroxy ketones give a positive reaction. Insoluble compounds (such as aromatic aldehydes) may react only slowly.

(iv) *Test.* Add the unknown (10 mg) to Schiff's reagent and see if a pink colour develops.

Comment. Aldehydes restore the pink colour to this reagent. Insoluble compounds may react only slowly.

(v) *Iodoform Test.* Mix the unknown (10 mg) with water (1 ml) and iodine solution (0.3 ml)* and then add sodium hydroxide solution dropwise and with shaking until the solution is pale yellow or colourless. Observe whether iodoform appears in the cold or on gentle warming. Iodoform appears as yellow crystals with characteristic smell.

If the unknown is not soluble in water this reaction can be carried out in dioxan or aqueous dioxan. The reaction mixture is finally diluted with water (2 volumes) to precipitate the iodoform. Since dioxan may contain impurities which give a positive reaction, carry out a blank experiment (without the unknown compound) at the same time.

Comment. The iodoform reaction is given by methyl ketones ($R{\cdot}CO{\cdot}CH_3$) and by compounds, containing the group $R{\cdot}CH(OH){\cdot}CH_3$, oxidizable to methyl ketones.

9c. Test for esters

To the unknown compound (1 drop) add saturated alcoholic hydroxylamine hydrochloride (3 drops) and 20% methanolic potassium hydroxide (3 drops). Heat the mixture until it boils, acidify with hydrochloric acid (0.5 *M*), and add ferric chloride solution dropwise. Look out for a deep red or purple coloration produced by the ferric salt of the hydroxamic acid.

Comment. Hydroxamic acids are produced from esters (and also from acid chlorides and anhydrides) as shown:

$$R{\cdot}C\begin{matrix}\nearrow O \\ \searrow OR'\end{matrix} \xrightarrow[KOH]{NH_2OH} R{\cdot}C\begin{matrix}\nearrow O \\ \searrow NHOH\end{matrix} \rightleftharpoons R{\cdot}C\begin{matrix}\nearrow OH \\ \searrow NOH\end{matrix}$$

* A solution of potassium iodide (20 g) and iodine (10 g) in water (100 ml).

Other compounds which react with ferric chloride [acids, phenols, enols—see Test 7(ii)] may also give colours and interfere with this test.

10. Alcohols

Strong evidence for an OH group (not necessarily alcoholic) should be available from the infrared spectrum. The following tests for alcoholic hydroxyl groups are valid only for compounds which do not contain nitrogen and are not acids, phenols, or carbonyl compounds. (Although phenols share many of the properties of alcohols, they are better examined as weak acids, Test 7.) These tests are also vitiated by the presence of water so the unknown compound must be dry. If it is a liquid, stand it over potassium carbonate for a few minutes and decant into a dry tube.

Test. (i) Add a small piece of freshly cut sodium to the liquid (3-5 drops) and observe whether any hydrogen is liberated. If there is no reaction, warm gently to about 80°, but disregard any slight effervescence which quickly slackens.* Solids can be tested after being dissolved in dry benzene.

(ii) Add acetyl chloride (2 drops) to a similar amount of unknown and observe whether any heat is developed or whether hydrogen chloride is evolved.

(iii) If these tests are positive apply the iodoform test [9b(v)].

Comment. Positive tests indicate an alcoholic group only if the groups listed above are absent. The last test is positive only for alcohols containing the group $CH_3 \cdot CH(OH)$– (and for methyl ketones).

11. Nitrogen-containing compounds

Sections 11-14 are concerned with nitrogen-containing compounds. The presence of this element should have been discovered in Test 4. In addition, amines are basic compounds (Test 6) and primary and secondary amines show characteristic infrared absorption. Similar absorption is shown by amides ($R \cdot CONH_2$) and substituted amides ($R \cdot CONHR'$) but these also have a carbonyl band and are neutral. Nitro compounds are also neutral and have characteristic infrared spectra; they are often

* Unreacted sodium should be destroyed by reaction with cold methanol or ethanol. When all the sodium has reacted (and not before) the solution can be washed down the sink.

coloured and may exhibit changes in colour with change of pH (see section 2). Remember that the unknown may contain more than one type of nitrogen-containing group.

12. Amines

(Read section 11 before carrying out any of these tests. See also p. 96 on the classification of amines.) The reaction between amines and nitrous acid provides useful information about the type of amine group. Whilst it is desirable to distinguish primary from secondary amines, it is imperative to distinguish these two classes from tertiary amines because the characterizing derivatives of primary and secondary amines are the same, but different from those used with tertiary amines. If there is any doubt about the result at any stage of this test, repeat it with an authentic sample of an amine of appropriate type.

In the reaction with nitrous acid the various amines react as follows:

	Aliphatic	*Aromatic*
Primary	Nitrogen liberated, alcohol (and other compounds) formed	Diazonium salt formed; gives azo dye with phenols
Secondary	Insoluble *N*-nitrosoamine formed	Insoluble *N*-nitrosoamine formed
Tertiary	No visible reaction	Gives *C*-nitroso compound which is usually green.

Test. (i) Undesirable side-reactions may occur unless these directions are followed carefully. Dissolve the amine (2 drops) in dilute hydrochloric acid (5-10 drops) with warming if necessary and cool the solution to 0-5° in an ice bath. If a solid separates, shake vigorously to produce fine crystals. Prepare a solution of sodium nitrite (50-100 mg) in water (0.2 ml) and add this dropwise to the amine solution with continuous shaking. It is important that the temperature should not rise above 5° during this operation. Watch out for three possible results.

Comment. (*a*) Steady evolution of nitrogen indicates the presence of a primary aliphatic amine. Take care not to confuse this with the slow decomposition of nitrous acid.

(*b*) The appearance of a precipitate (usually a liquid but sometimes a solid) indicates a secondary amine (aliphatic or aromatic). Since aromatic tertiary amines sometimes give a brown oil or precipitate at this stage, basify the solution. If a tertiary aromatic amine is present, the solution will turn green and a precipitate (solid or liquid) should appear.

(*c*) If the reactions described in (*a*) and (*b*) are not observed, divide the solution into two parts. Add one portion to a solution of β-naphthol (50 mg) in dilute sodium hydroxide (0.5-1.0 ml). An orange or red precipitate will be formed from a primary aromatic amine. If this test is negative add alkali to the other portion. If a tertiary aromatic amine is present, the solution will turn blue or green and a solid or liquid precipitate should appear.

Test. (ii) Add the unknown (1 drop or 50 mg) to bromine water (1 ml).

Comment. Many aromatic amines (and phenols) give a precipitate of polybromo-compound when treated with bromine water. Since bromine water is usually a very weak reagent take care not to add too much of the organic compound.

13. Nitro group

(Read section 11 before carrying out this test.) There follows a useful test for nitro compounds. Preparative reduction to amines is best effected with tin and hydrochloric acid (see section 35), hydrogen and a catalyst, or hydrazine hydrate and Raney nickel.

Test. Add the unknown compound (1 drop or 20 mg) to about 15 drops of titanous chloride solution which should be purple. If necessary, add just enough acetone to give a homogeneous solution. Observe if the purple colour is discharged within two minutes. The mixture may be heated on the water bath.

Comment. During this reaction a purple titanous salt is oxidized to a colourless titanic salt. Besides nitro compounds, nitroso-, azoxy-, azo-, and hydrazo-compounds, alkyl nitrates and nitrites, substituted hydroxylamines, and quinones also give a positive reaction.

14. Other nitrogen-containing compounds

(Read section 11 before carrying out this test.) These may include amides, nitriles, and compounds containing $>C=N-$(e.g. oximes). In order to study these, the compounds must be hydrolysed by acid or alkali and the products isolated and identified. For details of hydrolysis see section 34.

15. Compounds containing sulphur

If the compound does not also contain nitrogen it is probably a sulphonic acid or its salt, a sulphone (RSO_2R', very inert), thiol (RSH, very unpleasant smell), thio-ether (RSR′), sulphoxide (RSOR′), or sulphonyl halide (RSO_2Cl).

If nitrogen is also present it may be a sulphonamide, an amine sulphate, aminosulphonic acid, thiourea, or a dyestuff containing a sulphonic acid. Sulphonamides of primary amines (RSO_2NHR') are soluble in alkali. Amino-sulphonic acids are usually insoluble in ether and sparingly soluble in cold water; they react with bicarbonate solution but do not dissolve in dilute hydrochloric acid. Thiourea is soluble in water and sparingly soluble in ether; when fused it forms ammonium thiocyanate which gives a red colour with solutions containing ferric ions. Aryl substituted thioureas are soluble in sodium hydroxide.

16. Compounds containing halogen

These are acyl halides, alkyl halides, aryl halides, or salts containing a halide ion.

(i) *Test.* Shake the compound with aqueous silver nitrate and observe whether any silver halide is precipitated.

Comment. Salts containing halide ion react immediately. Acyl halides also give a positive reaction since these are decomposed by water (some slowly, some readily) to furnish halide ions. Some other halides such as t-alkyl and allyl halides are also hydrolysed quickly enough to give a positive test.

(ii) *Test.* Observe whether the substance fumes in air and reacts vigorously with water.

Comment. Some acyl halides react in this way.

(iii) *Test.* Boil with alcoholic potassium hydroxide, acidify with dilute nitric acid (5 *M*), and test again with aqueous silver nitrate.

Comment. Most alkyl halides, but not aryl halides, react sufficiently to give a positive halide test.

17. Compounds in which no functional group has been detected

If the preliminary tests do not indicate any functional group, and nitrogen, sulphur, and halogen are absent, test for unsaturation (Test 18). If this is absent the compound is probably a saturated hydrocarbon or ether and may be aromatic or aliphatic. The following tests should be applied only when no other positive reaction has been obtained, and even then only with caution in case some important observation has been overlooked.

17a. Solubility in cold concentrated sulphuric acid

Observe whether the unknown substance (1 drop or 20 mg) dissolves in cold concentrated sulphuric acid (5-10 drops).

Comment. Ethers and a few reactive aromatic hydrocarbons dissolve. Aliphatic and most aromatic hydrocarbons are insoluble. The ethers are recovered unchanged on dilution with water.

17b. Test with nitrating mixture

(*Apply this test only to those compounds whose infrared spectra show them to be aromatic.*) Shake the unknown compound (2 drops) with a mixture of concentrated nitric acid (5 drops) and sulphuric acid (5 drops). Observe any heat of reaction. If the result is in doubt, warm the mixture to 60° for 5 minutes, cool, add water (2 ml), recover any insoluble material, wash it with water, and test for the presence of a nitro group (infrared spectrum and Test 13).

Comment. Aromatic compounds react with production of heat; the corresponding aliphatic compounds are usually unaffected.

18. Test for unsaturation

The following tests for olefinic and acetylenic compounds are of limited value because other types of compounds sometimes give positive reactions, and because compounds of low solubility sometimes react only slowly.

(i) *Test.* Dissolve the unknown (1 drop) in carbon tetrachloride (0.5-1.0 ml) and add dropwise a 1% solution of bromine in this same solvent. Observe whether the colour is discharged.

Comment. Unsaturation is indicated if the bromine colour is discharged without evolution of hydrogen bromide. Active aromatic compounds (e.g. amines and phenols) may react by substitution but hydrogen bromide is generated.

(ii) *Test.* Dissolve the unknown (1 drop) in water or acetone (0.5-1.0 ml). Add 1% aqueous potassium permanganate solution dropwise and observe any colour change.

Comment. Immediate decolorization of an appreciable volume of reagent indicates unsaturation. Other easily-oxidized compounds (aldehydes, alcohols, aromatic amines and phenols) may react, but usually more slowly.

19. Preparation of characterizing derivatives

Once the functional groups present in an unknown compound have been recognized, the final stage of complete identification can be undertaken. Solids may be recognized by their melting point and their identity confirmed by mixed melting point, identity of infrared spectra, or chromatographic (GLC) behaviour. Liquids may also be recognized by the identity of their infrared spectra or GLC behaviour with that of an authentic sample. In many cases, however, this will not be possible and the unknown has to be converted into a derivative suitable for melting-point determination or chromatographic examination. The choice of derivative will depend on the functional groups which are present.

There follows a list of suitable derivatives and instructions for making these. Since a derivative is hardly ever pure when first prepared, it must be recrystallized before its melting point has any significant value. This may be less necessary for GLC exami-

Derivative	Reagents	Section*
Carboxylic acid, sulphonic acid, acid chloride, acid anhydride		
amide	ammonia	20
anilide	aniline	20
p-toluidide	*p*-toluidine	20
†methyl ester	methanol and hydrogen chloride or sulphuric acid; methanol and boron trifluoride	21
p-nitrobenzyl ester	*p*-nitrobenzyl bromide	22
S-benzylisothiouronium salt	*S*-benzylisothiouronium chloride	23
Aldehyde, ketone		
2,4-dinitrophenylhydrazone	2,4-dinitrophenylhydrazine	24
semicarbazone	semicarbazide hydrochloride	24
oxime	hydroxylamine hydrochloride	24
anil (aromatic aldehydes only)	aniline	24
Alcohol		
†acetate	acetic anhydride	26
†trimethylsilyl ether	hexamethyldisilazane and chlorotrimethylsilane or bis(trimethylsilyl)acetamide	30
3,5-dinitrobenzoate	3,5-dinitrobenzoyl chloride	27
Phenol		
†acetate	acetic anhydride	26
†trimethylsilyl ether	hexamethyldisilazane and chlorotrimethylsilane or bis(trimethylsilyl)acetamide	30
3,5-dinitrobenzoate	3,5-dinitrobenzoyl chloride	27
benzoate	benzoyl chloride	25
benzenesulphonate	benzenesulphonyl chloride	25
toluene-*p*-sulphonate	toluene-*p*-sulphonyl chloride	25
aryloxyacetic acid	chloroacetic acid	28
2,4-dinitrophenyl ether	chloro-2,4-dinitrobenzene	29
Primary and secondary amine		
acetyl amine	acetic anhydride or acetyl chloride	26
benzoyl amine	benzoyl chloride	25
benzenesulphonyl amine	benzenesulphonyl chloride	25
toluene-*p*-sulphonyl amine	toluene-*p*-sulphonyl chloride	25
picrate	picric acid	32
2,4-dinitrophenyl amine	chloro-2,4-dinitrobenzene	29
Tertiary amine		
picrate	picric acid	32
methiodide	methyl iodide	31

Derivative	Reagents	Section*
Alkyl halide		
S-alkylisothiouronium picrate	thiourea and picric acid	33
Amide nitrile, ester		
hydrolyse, isolate, and identify the hydrolysis products (carboxylic or sulphonic acids, alcohols or phenols, amines)		34
Nitro compound		
reduction	tin and hydrochloric acid	35
nitration (if aromatic)	nitric acid and sulphuric acid	36
Aromatic hydrocarbon		
picrate	picric acid	32
nitro compound	nitric acid and sulphuric acid	36

* Instructions for preparing these derivatives are given in the sections indicated.
† Recommended for GLC examination.

nation but then it is essential (almost always) to have an authentic sample of the compound for comparison. The unknown and/or its derivative and the authentic sample should show the same retention characteristics on subsequent chromatograms or when mixed, preferably on more than one stationary phase (see p. 23).

20. Amides, anilides, and p-toluidides from carboxylic and sulphonic acids (and acyl halides and anhydrides)

Carboxylic acids are best converted to these derivatives via the acid chloride; acid anhydrides, like the acyl halides, react directly with ammonia and amines. The acid chloride is conveniently prepared by reaction with thionyl chloride, but where this is unsuitable phosphorus pentachloride may have to be used. The reaction between an acyl halide and ammonia or amine may be vigorous: operate with care in a fume cupboard.

Reflux the acid (50-100 mg) with excess of thionyl chloride (2 ml) for 30 minutes in a fume cupboard. Remove unreacted thionyl chloride by connecting the reaction flask to the water pump and swirling the flask while keeping it immersed in warm water. These conditions may have to be moderated if the acid chloride is volatile.

To prepare the *amide*, add the acid chloride dropwise to strong ammonia (*d* 0.880,

1-2 ml). When the reaction has subsided and the solution is cold, filter the amide, wash with a little water, and crystallize from ethanol, aqueous ethanol, or (occasionally) water.

To prepare the *anilide* or p-toluidide, add aniline (0.2 ml) or finely-powdered *p*-toluidine (0.2 g) to the acid chloride, shake, and after a few minutes add dilute hydrochloric acid to dissolve the excess of amine and stir thoroughly to break up any lumps of solid which may be present. Filter the crude product, wash it with cold water, and crystallize from an appropriate solvent (usually ethanol or aqueous ethanol).

Heat the dry *sulphonic acid* or its salt (100 mg) with phosphorus pentachloride (200 mg) at 100° for 15-30 minutes. Cool, add water (1-2 ml), and remove the insoluble sulphonyl chloride. Wash this with cold water, then react it with strong ammonia (1 ml), heating the mixture to 100°. If the crystals do not appear on cooling, reduce the volume of the solution.

21. Methyl esters of carboxylic acids

These are seldom suitable for characterization by melting point but can be used for gas liquid chromatography where they are preferred to the free acids.

(*a*) Reflux the carboxylic acid (50-100 mg) with methanol (0.5-1.0 ml) containing 1-3% of hydrogen chloride or conc. sulphuric acid for 10-60 minutes. Add water (5 ml), extract the ester with ether, and wash the ethereal extract with sodium hydroxide solution (2 x 1 ml) and water (2 x 1 ml). Dry the solution and remove all or most of the ether.

(*b*) Boil the carboxylic acid (50 mg) with methanolic boron trifluoride (3%, 10 ml) for two minutes. Add water (5 ml) and ether or petroleum (5 ml); shake and separate the organic extract. Dry this solution, then remove all or most of the solvent.

22. p-Nitrobenzyl esters of carboxylic acids

Make a solution or suspension of the acid (100 mg) in water (0.5 ml) alkaline to phenolphthalein by adding sodium hydroxide solution (2 *M*) dropwise. Neutralize the excess of alkali with a little of the carboxylic acid or with very dilute hydrochloric acid. Add this to a solution of *p*-nitrobenzyl bromide* (100 mg) in ethanol (1 ml)

* p-Nitrobenzyl bromide is a skin irritant. If it gets on the skin, wash the affected parts immediately with ethanol.

along with additional ethanol if required to make the solution homogeneous. After refluxing for one hour and cooling, the ester should separate and can be recrystallized from ethanol or aqueous ethanol.

Longer heating and additional reagent may be necessary with polybasic acids.

23. S-Benzylisothiouronium salts of carboxylic and sulphonic acids

Prepare a solution of the sodium salt of the acid (100 mg) as described in section 22 and add this to a solution of *S*-benzylisothiouronium chloride (0.4 g) in water (1 ml). Cool the mixture in an ice bath until the derivative separates. It can be recrystallized from water, aqueous ethanol, or ethanol.

24. Derivatives of aldehydes and ketones

(a) *2,4-Dinitrophenylhydrazones.* Add the carbonyl compound (50-100 mg, dissolved in the minimum quantity of ethanol if it is solid) to an acidic solution of 2,4-dinitrophenylhydrazine (0.2 *M* Brady's solution, 2-4 ml)* and shake the mixture. If precipitation does not occur after 5 minutes, heat on the water bath for a further 5 minutes and cool; if a precipitate still does not form, add a little dilute sulphuric acid. Crystallize from an appropriate solvent (ethanol or acetic acid if necessary).

(b) *Semicarbazones.* Dissolve semicarbazide hydrochloride (100 mg) and sodium acetate (150 mg) in a few drops of water (not more), and add the carbonyl compound (50-100 mg). If this does not dissolve, add a little ethanol: if a granular precipitate forms at this stage it may be sodium chloride. In many cases the semicarbazone will separate on brief standing; otherwise, heat on the water bath for 5-10 minutes and then cool. If ethanol was added, it may be necessary to dilute with water. Crystallize the semicarbazone from an appropriate solvent (methanol, ethanol, benzene, or acetic acid).

(c) *Oximes.* Aldehydes form oximes readily, usually when the reagents are mixed at room temperature. With ketones long standing or refluxing is generally necessary.

Dissolve the carbonyl compound (50-100 mg) in ethanol (0.5 ml) and mix it with a solution of hydroxylamine hydrochloride (50-100 mg) and sodium acetate (100-200

* See footnote, p. 74.

mg) in water (0.2-0.3 ml). Reflux the mixture for 5-10 minutes (somewhat longer for ketones) then cool and neutralize with hydrochloric acid. Crystallize the derivative from a suitable solvent (ethanol).

(d) *Anils of aromatic aldehydes.* Heat the aldehyde (100 mg) and aniline (100 mg) alone or dissolved in the minimum amount of warm ethanol, at 100° for 10 minutes, then cool in ice when solidification should occur. If this does not happen, heat the mixture a little longer. Recrystallize the anil from ethanol.

25. Benzoyl, benzenesulphonyl, and toluene-p-sulphonyl derivatives of phenols and of primary and secondary amines

Benzoyl derivatives are conveniently prepared by the Schotten-Baumann reaction, but for the sulphonyl derivatives the pyridine method is more satisfactory. The instructions are given for phenols, but amines react similarly except that they may not be soluble in aqueous alkali. The sulphonamides of primary amines are soluble in alkaline solution.

(a) *Schotten-Baumann reaction. The most common cause of failure of this reaction is the use of too much benzoyl chloride.* Take care not to exceed the stipulated quantities.*

Place the phenol (100 mg) in a tube (provided with a well-fitting stopper) along with 2 *M* sodium hydroxide solution (1.5 ml) and benzoyl chloride (3 x 0.05 ml). After addition of each portion of benzoyl chloride, shake well for 2-3 minutes and cool under the tap if the mixture gets very warm. Finally shake for 5-10 minutes when the smell of benzoyl chloride should have disappeared and the solution should still be alkaline (check this). If benzoyl chloride is still present, add more sodium hydroxide solution (1 ml), shake for a further 5 minutes, and repeat this process if necessary. A solid should separate at this stage; remove the liquid, wash the solid with water, and crystallize from an appropriate solvent (probably ethanol).

(b) *Pyridine method.* Heat a mixture of phenol (100 mg) and pyridine (0.5 ml) with benzoyl chloride* or benzenesulphonyl chloride or toluene-*p*-sulphonyl chloride (200

* Benzoyl chloride is an unpleasant lachrymatory chemical and should be handled with care. Never try to wash it away with *hot* water. If necessary, use strong ammonia solution to destroy it.

mg) on a water bath for 15 minutes, and then add water (3-5 ml). A solid should separate. Remove the liquid and wash the solid with dilute acid to remove pyridine (and amine if present), with dilute alkali to remove phenol and acid or acid chloride, and with water. Finally crystallize from a suitable solvent (probably ethanol). If the product appears to be an oil instead of a solid, it should be treated in the same way; it will probably crystallize as impurities are removed during washing.

26. Acetylation of amines, phenols and alcohols

The first method is suitable for the preparation of solid acetyl derivatives (mainly of amines) which are to be characterized by melting point. The second method is for liquid acetates which are to be examined by GLC.

(*a*) Dissolve the amine (50-100 mg) in glacial acetic acid (0.1-0.2 ml), add acetic anhydride (0.1-0.2 ml), and warm gently on a water bath for 5 minutes. Add water (1-2 ml) dropwise and with shaking to the cold reaction mixture (this reaction may be vigorous). If the product does not appear at this stage, basify the solution with strong ammonia. Crystallize from a suitable solvent (water, aqueous alcohol, or alcohol).

(*b*) Reflux the alcohol (50-100 mg) with acetic anhydride (0.5-1.0 ml). If the acetate is volatile, dilute the reaction mixture with water and extract with ether; if it is not volatile, remove the excess of anhydride by gentle heating under reduced pressure.

27. 3,5-Dinitrobenzoates of alcohols or phenols

The first method is suitable only for the simpler volatile alcohols; the second method is a more general procedure.

(*a*) Heat a mixture of 3,5-dinitrobenzoyl chloride (200 mg) and alcohol (100 mg) on a water bath for 10 minutes. The ester should crystallize on cooling or after concentration, and is crystallized from petroleum (b.p. 40-60°) or carbon tetrachloride.

(*b*) Reflux the alcohol or phenol (100 mg) with pyridine (0.5 ml), 3,5-dinitrobenzoyl chloride (200 mg), and dry benzene (2 ml) for 15 minutes (30 minutes for tertiary alcohols). Add ether (3-5 ml) to the cold solution and wash with dilute hydrochloric acid to remove pyridine, with dilute sodium hydroxide to remove 3,5-dinitrobenzoic acid and phenol, and with water. After removal of solvent, crystallize from aqueous alcohol, petroleum (b.p. 40-60°), or carbon tetrachloride.

28. Aryloxyacetic acids derived from phenols

Dissolve or suspend the phenol (50-100 mg) in sodium hydroxide solution (8 *M*, 0.2-0.3 ml) and add chloroacetic acid (60-120 mg) dissolved in water (0.2-0.3 ml). Warm the mixture on a steam bath for one hour, then cool and neutralize with hydrochloric acid. If the derivative does not separate as a solid, extract the solution with ether (5 ml). Wash this with water to remove chloroacetic acid and then with sodium carbonate solution (1 *M*) to remove the aryloxyacetic acid. Cautiously acidify this extract and crystallize the derivative from water or aqueous ethanol.

29. 2,4-Dinitrophenyl derivatives of phenols and of primary and secondary amines

(a) *Phenols.* Dissolve the phenol (100 mg) in strong potassium hydroxide (10 *M*, 1 ml) and heat with chloro-2,4-dinitrobenzene* (100 mg) at 100° for 1 hour. Dilute the mixture with water, filter off the solid, wash with dilute sodium hydroxide and with water, and recrystallize from ethanol.

(b) *Amines.* Heat the amine (100 mg) and chloro-2,4-dinitrobenzene* (100 mg) at 100° for 5-10 minutes. The mixture should solidify on cooling and the product can be crystallized from ethanol or acetic acid.

30. Trimethylsilyl derivatives of hydroxy compounds

The polar hydroxy compounds are more conveniently examined by GLC as their trimethylsilyl derivatives. Two methods are described.

(i) Dissolve the hydroxy compound (5-10 mg) in pyridine (0.5 ml) and add hexamethyldisilazane (0.2 ml) and trimethylchlorosilane (0.1 ml). Shake for 1 minute and allow to stand for 5 minutes. Add petroleum (b.p. 40-60°, 5 ml) and water (5 ml) and shake the mixture. Separate the petroleum extract, repeat with a second portion of petroleum (5 ml), and dry the combined extracts with anhydrous sodium sulphate. Distil off most of the solvent and use the residue for GLC examination.

(ii) Add bis(trimethylsilyl)acetamide (0.2-0.3 ml) to the alcohol (2-3 mg) in a

* Chlorodinitrobenzene is a skin irritant. If it gets on the skin, wash the affected parts immediately with ethanol.

test tube, stopper the tube, and shake for 1 minute. The mixture is injected directly on to the GLC column.

31. Methiodides of tertiary amines

Mix the amine (50-100 mg) with methyl iodide (0.1-0.2 ml) and keep at room temperature for 5 minutes and then at 60° for 5 minutes. Cool in ice and scratch if necessary to induce crystallization. Recrystallize from an appropriate solvent (ethanol or acetone). These compounds are deliquescent and are difficult to crystallize in the presence of water.

32. Picrates of amines and of aromatic hydrocarbons

Many aromatic compounds give crystalline molecular complexes with picric acid; the amines react to form salts. Mix strong warm solutions of the amine or hydrocarbon (50-100 mg) and of picric acid (50-100 mg). For amines use benzene as solvent, for other aromatic compounds use ethanol or benzene. The product should separate on cooling and can be crystallized from the original solvent.

33. S-Alkylisothiouronium picrates

These are suitable derivatives for primary and secondary alkyl bromides and iodides. Chlorides react only slowly and the reaction is not suitable for tertiary halides.

Reflux a mixture of powdered thiourea (100 mg), alkyl halide (100 mg), and ethanol (1 ml) for the appropriate time (see below). Then add picric acid (100 mg), boil until the solution is clear, and cool. If no precipitate forms, add a little water. Recrystallize the picrate from ethanol.

Reaction time: primary bromides and iodides, 20 minutes; secondary bromides and iodides, 3 hours; alkyl chlorides, 5 hours.

34. Hydrolysis of esters, amides, and nitriles

(i) Esters are hydrolysed by refluxing with aqueous or alcoholic solutions of potassium hydroxide (5 *M*). The reaction may be complete in 30 minutes or may require several hours.

(ii) Amides are more conveniently hydrolysed under acidic conditions. Boil with aqueous sulphuric acid (9 *M*) or with an ethanolic solution of concentrated hydrochloric acid.

(iii) Nitriles are sometimes hydrolysed by prolonged reaction as in (i) or (ii). Alternatively use potassium hydroxide in 2-methoxyethanol (5 *M*) or a mixture of acetic acid, water, and concentrated sulphuric acid (1:1:1).

35. Reduction of nitro compounds

Suspend the nitro compound (200 mg) in concentrated hydrochloric acid (5 ml) and add a few small granules of tin (*ca.* 1 g). After heating for 30 minutes decant from unreacted tin into excess of sodium hydroxide solution (8 *M*) in which the initially precipitated tin salts should redissolve. The amine is recovered by steam distillation or extraction or filtration as appropriate. If solid, crystallize from benzene or petroleum; if liquid, convert the amine into some solid derivative.

36. Nitration

Aromatic compounds vary considerably in the ease with which they undergo nitration. Several methods are therefore suggested and these are given below in order of increasing vigour. *Each of them should be carried out with great care.*

(i) Add the substance (100 mg) gradually to a mixture of concentrated nitric acid (1 ml) and sulphuric acid (1 ml), keeping the temperature below 25°. Shake gently until reaction appears to be complete. If there is no sign of reaction, warm gently to about 40°. Reaction is complete when a test portion solidifies when poured into water. Pour the reaction mixture into water and crystallize the nitro compound from a suitable solvent (ethanol).

(ii) Add fuming nitric acid (1 ml) dropwise and with shaking to the substance (100 mg) cooled in ice-water. Allow the reaction to subside between each addition. After standing at room temperature for 5 minutes, pour into water and work up as in (i).

(iii) Carefully mix fuming nitric acid (0.6 ml) and concentrated sulphuric acid (1 ml) and to this mixture at room temperature gradually add the substance to be nitrated (100 mg). Shake the mixture frequently and cool if necessary. When the reaction finally subsides, heat on the steam bath for 5 minutes then pour into water and recover as in (i).

Note on Experiment 7

Alkenes should decolorize the bromine solution and give a brown precipitate with the permanganate solution. Alkenes may fail to undergo these reactions because the double bond is markedly deactivated by adjacent electron-withdrawing groups or because of their low solubility in the reaction solvent.

Some compounds which are not alkenes give positive reactions in one or both of these tests. Any easily oxidizable compound, such as an aldehyde, phenol, or aromatic amine, will react with potassium permanganate. Compounds with highly activated aromatic systems (phenols and amines) react with bromine in a substitution reaction and the bromine solution is thereby decolorized.

Index